T0249587

Artificial Vision
Image Description, Recognition and Communication

Signal Processing and its Applications

SERIES EDITORS

Dr Richard Green
The Engineering Practice
Farnborough, UK

Professor Doug Gray
CRC for Sensor Signal and Information Processing, Signal Processing Research Institute,
Technology Park, Adelaide, South Australia

Professor Edward J. Powers
Department of Electrical and Computer Engineering,
University of Texas at Austin, USA

EDITORIAL BOARD

Professor Maurice G. Bellanger
CNAM, Paris, France

Professor Gerry D. Cain
School of Electronic and Manufacturing Systems Engineering,
University of Westminster, London, UK

Dr Mark Sandler
Department of Electronics and Electrical Engineering,
King's College London, University of London, UK

Professor Colin Cowan
Department of Electronics and Electrical Engineering,
Loughborough University of Technology, Loughborough, UK

Dr David Bull
Department of Electrical Engineering,
University of Bristol, UK

Professor D.J. Creasey (deceased)
School of Electronic and Electrical Engineering,
University of Birmingham, UK

Dr Henry Stark
Electrical and Computer Engineering Department,
Illinois Institute of Technology, Chicago, USA

Artificial Vision

Image Description, Recognition and Communication

edited by

VIRGINIO CANTONI
Dipartimento di Informatica e Sistemistica
Università di Pavia
Via Ferrata 1
27100 Pavia, Italy

STEFANO LEVIALDI
Dipartimento di Scienze dell'Informazione
Università di Roma "La Sapienza"
Via Salaria 113
00198 Roma, Italy

VITO ROBERTO
Dipartimento di Informatica
Università di Udine
Via delle Scienze 206
33100 Udine, Italy

ACADEMIC PRESS
San Diego London Boston New York Sydney Tokyo Toronto

This book is printed on acid-free paper.

Copyright © 1997 by ACADEMIC PRESS

All Rights Reserved
No part of this publication may be reproduced or transmitted in any form or by any means,
electronic or mechanical, including photocopy, recording, or any information storage and
retrieval system, without permission in writing from the publisher.

Academic Press, Inc.
525 B Street, Suite 1900, San Diego, California 92101–4495, USA

Academic Press Limited
24–28 Oval Road, London NW1 7DX, UK

ISBN 0-12-444816-X

A catalogue record for this book is available from the British Library

Printed and bound by CPI Antony Rowe, Eastbourne

Series Preface

Signal processing applications are now widespread. Relatively cheap consumer products through to the more expensive military and industrial systems extensively exploit this technology. This spread was initiated in the 1960s by introduction of cheap digital technology to implement signal processing algorithms in real-time for some applications. Since that time semiconductor technology has developed rapidly to support the spread. In parallel, an ever increasing body of mathematical theory is being used to develop signal processing algorithms. The basic mathematical foundations, however, have been known and well understood for some time.

Signal Processing and its Applications addresses the entire breadth and depth of the subject with texts that cover the theory, technology and applications of signal processing in its widest sense. This is reflected in the composition of the Editorial Board, who have interests in:

(i) Theory – The physics of the application and the mathematics to model the system;

(ii) Implementation – VLSI/ASIC design, computer architecture, numerical methods, systems design methodology, and CAE;

(iii) Applications – Speech, sonar, radar, seismic, medical, communications (both audio and video), guidance, navigation, remote sensing, imaging, survey, archiving, non-destructive and non-intrusive testing, and personal entertainment.

Signal Processing and its Applications will typically be of most interest to postgraduate students, academics, and practising engineers who work in the field and develop signal processing applications. Some texts may also be of interest to final year undergraduates.

Richard C. Green
The Engineering Practice,
Farnborough, UK

About the Editors

Virginio Cantonio is presently Full Professor of Computer Programming at Pavia University. His most recent work is concerned with object recognition and parallel architectures for image processing and computer vision. He has been the coordinator of an Italian National Project involving a consortium of seven universities for the design and construction of a pyramidal system for image analysis. He is author or co-author of more than 120 journal or conference papers and book chapters and has edited or co-edited 12 books while being co-author of *Pyramidal Architectures for Computer Vision.*

Stefano Levialdi, Full Professor of Computer Science at Rome University, is the Director of the Pictorial Computing Laboratory. He has published over 180 scientific papers and edited over 20 books on image processing. He is founder and co-editor of the *Journal of Visual Languages and Computing* and Associate Editor of *Pattern Recognition, Computer Vision and Image Understanding, Pattern Recognition Letters, Machine Vision and Applications* and *Signal Processing.*

Vito Roberto is Associate Professor at the Computer Science Faculty, University of Udine, Italy. His research interests concern the design of intelligent systems for image interpretation and communication, in several application domains. He is responsible for the Machine Vision Laboratory at the same University. Professor Roberto is author of several articles and editor of many volumes in the field of pattern recognition and artificial intelligence.

Contents

List of Contributors

Ardizzone, E.
Dipartimento di Ingegneria Elettrica
Università di Palermo
Via delle Scienze
90128 Palermo, Italy

Bertolotto, M.
Dipartimento di Informatica e Scienze dell'Informazione
Università di Genova
Viale Benedetto XV, 3
16132 Genova, Italy

Bianchi, N.
Dipartimento di Fisica
Università di Milano
Via Viotti 5
20133 Milano, Italy

Bottoni, P.
Dipartimento di Scienze dell'Informazione
Università di Roma 'La Sapienza'
Via Salaria, 113
00198 Roma, Italy

Bruzzone, E.
Dipartimento di Informatica e Scienze dell'Informazione
Università di Genova
Viale Benedetto XV, 3
16132 Genova, Italy

Cantoni, V.
Dipartimento di Informatica e Sistemistica
Università di Pavia
Via Ferrata 1
27100 Pavia, Italy

Caputo, G.
Istituto di Psicologia
Università di Pavia
Piazza Botta 6
27100 Pavia, Italy

Chang, S.K.
Visual Computer Laboratory
Department of Computer Science
University of Pittsburgh
Pittsburgh, PA 15260
USA

Chella, A.
Dipartimento di Ingegneria Elettrica
Università di Palermo
Via delle Scienze
90128 Palermo, Italy

De Floriani, L.
Dipartimento di Informatica e Scienze dell'Informazione
Università di Genova
Viale Benedetto XV, 3
16132 Genova, Italy

Del Bimbo, A.
Dipartimento di Sistemi e Informatica
Università di Firenze
Via di Santa Marta, 3
50139 Firenze, Italy

Di Gesù, V.
Dipartimento di Matematica ed Applicazioni
Università di Palermo
Via Archirafi, 34
90124 Palermo, Italy

Gaglio, S.
Dipartimento di Ingegneria Elettrica
Università di Palermo
Via delle Scienze
90128 Palermo, Italy

Iacono, A.M.
Dipartimento di Filosofia
Università di Pisa
Italy

Levialdi, S.
Dipartimento di Scienze dell'Informazione
Università di Roma 'La Sapienza'
Via Salaria, 113
00198 Roma, Italy

Lombardi, L.
Dipartimento di Informatica e Sistemistica
Università di Pavia
Via Ferrata 1
27100 Pavia, Italy

Mussio, P.
Dipartimento di Scienze dell'Informazione
Università di Roma 'La Sapienza'
Via Salaria, 113
00198 Roma, Italy

Roberto, V.
Machine Vision Laboratory
Dipartimento di Informatica
Università di Udine
Via delle Scienze, 206
I 33100 Udine, Italy

Savini, M.
Dipartimento di Informatica e Sistemistica
Università di Pavia
Via Ferrata 1
27100 Pavia, Italy

Tegolo, D.
Dipartimento di Matematica ed Applicazioni
Università di Palermo
Via Archirafi, 34
90124 Palermo, Italy

Trucco, E.
Department of Computing and Electrical Engineering
Heriot-Watt University
Riccarton
Edinburgh EH14 4AS
Scotland

Verri, A.
INFM – Dipartimento di Fisica
Università di Genova
Via Dodecaneso, 33
16146 Genova, Italy

Yeshurun, Y.
Computer Science Department
School of Mathematical Sciences
Tel Aviv University
IL 69978 Tel Aviv
Israel

Preface

Artificial vision is a rapidly growing discipline, aiming to build computational models of the visual functionalities in humans, as well as machines that emulate them. Although the ultimate goals are ambitious, successful theories and applications have appeared so far in the literature, thanks to the converging contributions from image processing, computational geometry, optics, pattern recognition and basic computer science.

The on-going research work suggests that vision is to be studied in the context of other – seemingly distinct – functionalities. In particular, active vision focuses on the dynamics of sensing, so that moving, modifying sensor parameters, and acquiring information concur to perform visual tasks in a purposive way. Moreover, quite often visual modules are to be included into more complex 'intelligent' systems devoted to real-world applications – e.g., automated control, inspection and diagnosis. Models of the objects are to be acquired and stored; plausible hypotheses are to be maintained about the scene content. In this way, profound links emerge between visual and cognitive capabilities, like learning, reasoning and communicating. Visual communication, in itself, involves a number of challenging topics, with a dramatic impact on the contemporary culture, where human–computer interaction and human dialogue via computers play a more and more significant role.

The present volume contains a number of selected review articles concerning the research trends mentioned above. In particular, Part I groups contributions in active vision, Part II deals with the integration of visual with cognitive capabilities, while Part III concerns visual communication.

We hope that the topics reported in this volume will encourage further research work along the emerging directions, and towards an integrated, comprehensive study of vision and intelligence in humans and machines.

Part of the material has been presented at the fourth School on Machine Vision, organized by the Italian Chapter of the International Association for Pattern Recognition (IAPR), and held in Udine, Italy, 24–28 October 1994.

We wish to thank all the researchers who contributed to the success of the school and the preparation of this volume. In particular, we thank all the members of the scientific committee: L. Cordella (Napoli), S. Gaglio (Palermo),

S. Impedovo (Bari), G. Pieroni (Udine) and R. Stefanelli (Politecnico di Milano). The International Centre for Mechanical Sciences (CISM) is gratefully acknowledged for hosting the school. We are also grateful to Forum (Udine) for the help in typesetting and preparing the manuscripts.

Pavia, Rome and Udine, December 1995

Virginio Cantoni
Stefano Levialdi
Vito Roberto

Sponsoring Institutions

The following institutions are gratefully acknowledged for their support to the School on Machine Vision and the preparation of the present volume:

The Italian National Research Council (Consiglio Nazionale delle Ricerche – CNR), under grant no. AI95.01113.07.
The Friuli-Venezia Giulia Regional Government, grant no. 4943.
The Consorzio Universitario del Friuli, grant no. 60/94.

PART I

ACTIVE VISION

Real-time vision applications can achieve their required performances only by applying the appropriate computations on the relevant data, and at the right time. A 'smart' selection of data, located almost anywhere in the wide field of view, is generally performed at the early stages of the analysis, to broadly detect what is present in the scene, and where it is. The human camera-eye maneuvers for position and 'localizes' the regions of interest thanks to a sophisticated control. The highest concentration of sensing transducers occurs in the fovea, which occupies just a small portion of the field of view: about 120 minutes of arc around the visual axis; meanwhile, the sensing field is extended broadly, over 180 degrees. A complex 'alerting mechanism' allows one to effectively exploit the limited, but wealthy sensing resources of the fovea: as the scene changes, it rapidly guides the selection of the regions of interest to be focussed through the fovea. The selection itself requires eye vergence and head-and-eye movements, to scan the surrounding environment.

The following sections address the main aspects of the mentioned strategy.

The processes towards the area of scrutiny of an image, the general mechanisms that control the approaches, and the sequence of activities involved, are discussed in Chapter 1 from the point of view of biological systems; meanwhile, Chapter 2 considers the main features of attention mechanisms – and possible ways to model them – from the perspective of artificial systems. Chapter 3 emphasizes the purposeful modification of the parameters of the sensory apparatus, in order to simplify the visual tasks. Details are illustrated in Chapter 4 on the scan-path selection for model acquisition of unknown objects, and the solutions proposed in order to use the same models for planning sensor configurations in active vision. Finally, Chapter 5 deals with the basic regularization processes in early vision that drive the image analysis towards robust, goal-directed solutions.

1

Attentional Engagement in Vision Systems

1.1 INTRODUCTION

The evolution of technology in the last decades has been leading towards *active* visual sensors: the camera has evolved towards a visual front-end in which single or even multiple points of view are controlled on-line. This new equipment has fostered the development of new areas of research such as exploratory vision, a term introduced by Tsotsos (1994), in which the environment is analysed through a controllable camera exploiting investigations analogous to those supported by eye/head/body movements in human vision.

At a different level of control lie the sub-cases in which only the control system for camera orientation acts. Research in the field of biological and human vision, in particular, started several decades ago. The *scan paths* recorded by Yarbus (1967) are now well known. In the course of this work, the sequences of the fixation points of the observer's eye (when looking at paintings and other artifacts or at particular scenes of ordinary life) are first analysed. The order of the fixation points is not at all random, but the salient features of the visual field are inspected following a pathway with dense areas around the 'critical points'. The mean time between two fixation points is around 250 ms.

At the second level of the hierarchy, characterized by more subtle and

ARTIFICIAL VISION
ISBN 0-12-444816-X

Copyright © 1997 Academic Press Ltd
All rights of reproduction in any form reserved

detailed scans, is the element-by-element scrutiny carried out by the fast moving aperture of a *searchlight of attention* (with steps lasting 15–50 ms each). In this case no 'mechanical' movements are required, eyes are directed to some 'gross center of gravity' (Findlay, 1982) of points of interest, after which a searchlight scrutinizes each single point of interest step by step.

In this chapter the attention processes of the human visual system will be considered, outlining some of the analogies and differences with respect to the frameworks introduced by computer scientists in artificial vision. In the next section a review of psychological research on attention is presented, distinguishing the pre-attentive from the attentive theories. The former family groups the early vision models and the interrupt theories together; the latter includes selective and spatial attention theories. Section 1.3 deals with the main theories and outcomes of the research in neurosciences. Finally, a conjecture is briefly introduced regarding the functional characteristics and organization of the attention supervisor of high-level behavior.

1.2 ATTENTION IN PSYCHOLOGY

Scene understanding is quite a complex process in which many different objects, with different positions, under different lighting conditions, and from different perspectives, are gathered simultaneously. The combinatorial explosion of the possible instances of the same object precludes a discrete interpretation like the 'grandmother cell' concept in which there is a referent neuron for each possible instance (Barlow, 1972). Scene information coded by scene points in the retina whilst flowing through the visual system, is represented by simple pattern primitives in the striate cortex, and gathers into more and more complex structures in the extrastriate areas. Objects at the highest levels are encoded as *patterns of activation* in a complex network of information-transform neurons which are feature sensitive (Uhr, 1987; Van Essen *et al.*, 1992).

Going from the retinal to the central cortical neurons, the level of data abstraction gets higher, while precision in pattern location decreases. There is a semantic shift from 'where' towards 'what' in the scene (Burt *et al.*, 1986). Indeed, as information passes through the visual pathway, the growing receptive field of neurons produces a translation invariancy. At later stages, the large receptive fields of neurons lead to a mis-localization and the consequent mis-combination of features. Moreover, the simultaneous presence of multiple objects increases difficulties in determining which features belong to what objects. This difficulty has been termed the *binding problem* (Damasio, 1989; Engel *et al.*, 1992).

The binding problem may be solved by providing an *attentional mechanism* that only allows the information from a selected spatial location to act at higher stages (Treisman and Gelade, 1980; Luck and Hillyard, 1994a and b). From the computational view point, the early visual information processing is characterized by two functionally distinct modes (Neisser, 1967; Julesz, 1986):

- *pre-attentive mode* – the information is processed in a spatially parallel manner, without need of attentional resources. This parallel execution is characterized by: i) complete independence of the number of elements; ii) almost instantaneous execution; iii) lack of sophisticated scrutiny; iv) large visual field.
- *attentive mode* – for more complex analyses, the allocation of attentional resources to specific locations or objects is needed. A sequence of regions of interest is scrutinized by a small aperture of *focal attention*.

The object can be consciously identified only within this attentional aperture. Pre-attentive processes perform primitive analysis in order to segment or isolate areas of the visual field, and detect the potential regions of interest to which attention should be subsequently allocated for more sophisticated, resource-demanding processing. Attentional dispatching can be voluntary, i.e. guided by strategic goals, or involuntary, i.e. reacting to particular stimuli. These two spatial attention modes have been named *endogenous* and *exogenous*, respectively (Jonides, 1981).

1.2.1 Pre-attentive theories

A number of different models have been proposed to describe pre-attentive behavior. A first family of models includes the set of early-vision primitives belonging to the *reconstructionist approach* (Marr, 1982: see also AA.VV., 1994). A second family includes the reacting models capturing attention by detecting discontinuities and anomalies in space and/or in time. In the sequel several proposals are briefly reviewed.

1.2.1.1 *Early vision models*

Early vision, in the sense given by Marr, is a data-driven process in which sets of *intrinsic images* are derived from the visual scene to support a 3D reconstruction of the scene itself. Early processing is organized into a hierarchy of increasingly complex feature detectors (Hubel and Wiesel, 1962). The set of pre-attentive primitives is potentially wide and varies substantially with the context. In particular, it usually includes geometrical and physical entities such as color, morphology and motion.

Research conducted on both human and machine vision has led towards a framework called reconstructionism. This framework is based on the quoted physical properties, and holds that quantitative feature analysis (such as slant, tilt, reflectance, depth and motion) is a first step towards qualitative judgments

Figure 1.1 Three cases of regions composed by two different textures are shown. In the left region the segment on the left is composed by a letter L, the one on the right is composed by the letter T. The middle region contains L and T letters distributed randomly, but the left segment has a lower grey level than the right one. Finally, the right region contains L and T letters distributed randomly, both letters can be black or with an intermediate grey level: on the left segment, the Ts are black and the Ls are brighter; on the right segment, vice versa. The partition of the pairs of textures is quite evident in the left and middle regions and not detectable in the right one: shape and grey level produce pop-out, but their conjunctions do not!

and actions. In this respect, the most discussed paradigm in psychology is the *pop-out theory* introduced in the next section.

1.2.1.1.1 The pop-out theory. In Treisman and Gormican (1988) is defined a dimension as a set of mutually exclusive values for any single *feature stimulus*. For example, a line can be both red and vertical (values on two different dimensions: one corresponding to feature color and the other to feature orientation); obviously, a line cannot be both vertical and horizontal (values on the same dimension). According to Treisman, a feature is independent of any other instances that are present in the field; it is a distinct and separate entity.

The phenomenon in which targets characterized by a single feature, not shared by the other elements (usually called distractors) in the scene, are detected is called *pop-out*. In these cases the subject can see only a blur until suddenly the target jumps out before him. From the experimental point of view the response time[1] (RT) for target presence is independent of the number of elements (*flat search function*).

The early stages of visual processing extract a number of primitive features automatically and in parallel across the entire visual field. Possible primitive features of the early vision stage are: color and different levels of contrast, line curvature, acute angle, line tilt or misalignment, terminator and closure, size (Treisman and Gormican, 1988), direction of movement (McLeod *et al.*, 1991), stereoscopic disparity (Nakayama and Silverman, 1986), shading (Kleffner and Ramahandran, 1992), 3D interpretation of 2D shapes (Enns and Resnik, 1991),

---

[1] The experimental set-up considers the number of elements as the independent variable and the response time from stimulus onset the dependent parameter. Three classes of behavior were singled out: *flat search function* (the response time is independent of the number of elements); *negative search function* (the higher the number of elements the lower the response time); *positive search function* (the higher the number of elements the higher the response time) [Bravo and Nakayama, 1992].

etc. None of the following features, instead, meet the parallel search criterion: line arrangements like intersection and junction, topological properties like connectedness and containment, relational properties like height-to-width ratio (Treisman and Gormican, 1988), etc.

Some qualitative features seem to exhibit asymmetrical performance, e.g. when detecting curvature, tilt, elongation and convergence (Treisman and Gormican, 1988), cases in which one direction brings faster target detection than another. This implies that phenomenologically unitary features may be decomposed into a standard or reference value, by adding a value to signal any deviation. For example, a slightly tilted line may be detected as vertical plus some activity in diagonal feature detectors. So, asymmetry in tilt – a tilted line among vertical lines pops out, whereas the inverse yields to an attentive search – could be explained by the fact that a vertical line among tilted ones lacks the diagonal component.

From the computational viewpoint this phenomenon has been interpreted as if the targets with a single feature not shared by the distractors are checked for the presence/absence of the unique target feature by a parallel check in a specialized *feature map*. The subject checks a pooled response for the *presence* of activity anywhere in the map. As mentioned in the previous section, features are represented independently of their location: the observers can identify a feature without knowing where it is. Features are free floating, in the sense that without focused attention their locations are uncertain.

The pre-attentive search is a search in which attention is *distributed* (*diffused*, for some authors) widely over the visual field rather than narrowly *focused* and directed serially to an object or subgroup at a time (Treisman, 1991).

This theory justifies the aphorism 'presence is parallel, absence is serial'. While there is a pooled response for the presence of a feature not shared by distractors, a target is pre-attentively invisible if it is defined only by the *absence* of a feature that is present in all the distractors. If we measure the pooled response to the relevant feature, we can expect to see a difference between scenes

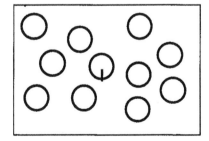 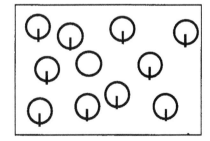

Figure 1.2 'Presence is parallel, absence is serial': the left region contains several circles, one of them with a crossing vertical segment; the right region vice versa contains just one plain circle, all the others having the crossing segment. The odd item is quite evident on the left: on the right a serial scrutiny is required.

containing $n - 1$ instances and scenes containing n instances. Soon the difference becomes unreliable given system noise, and subjects are forced to search serially.

1.2.1.1.2 ROI indexing. There is an intermediate stage between pre-attentive and attentive analysis: the *individuation* stage (Pylyshyn, 1989). A small number of salient items in the visual field are indexed and thereby made readily accessible for a variety of visual tasks. *Distal* features (which are currently projected onto the *proximal* features in the retina) can be referred to an indexing mechanism, and the information can be accessed directly from scene data. Data access can be achieved not by 'internalizing' images of the scene, but by leaving the information where it is in the world and gathering it indirectly only when necessary.

There should be a number of disparate loci selected by the early pre-attentive visual system that are potentially available in parallel. Indexing is the way operators or processes refer to interesting data. The indexes allow primitive visual entities to be accessed or interrogated to ROI through a pre-attentive binding process. This mechanism operates without an absolute coordinate system and with no need to refer to a particular feature.

The indexes can be assigned automatically (e.g. by abrupt onset) or be voluntary. These indexes do not need to be assigned to contiguous items or even to nearby items (this position differs from the concept of a unitary focus of attention), but can bind several disparate items scattered throughout the visual field. The selected items are accessed without requiring a search on the scene (this differs from the view that attention has to scan continuously from place to place and/or zoom). Moreover, an index is sticky: once assigned, it tends to remain with the item even when it moves around.

Indexing precedes attention, the features to be bound together have to be localized in advance to avoid illusory conjunctions (Treisman and Schmidt, 1982). Spatial relations between elements such as inside, connected, ahead of (Ullman, 1984) require the elements themselves to be indexed. It becomes necessary to move the focus of attention to the correct retinal location, finding a way to know where are the selected items.

Indexing is a parallel but resource-limited process: there are limitations on the number of items in the index set. An important and impressive example is *subitizing*, i.e., the capability to coarsely enumerate the number of singletons[2] in the field of view. If the number of singletons is limited (around four, according to most of the authors) the RT presents an almost flat slope and a high accuracy. When the number of items is higher, RT presents quite a prominent positive slope and accuracy is limited (Trick and Pylyshyn, 1993).

In other experiments, indexing has been used to explain the limits to visual tracking. In a typical study, ten or more identical disks move randomly around

---

[2] Target differing from distractors by at least one unique feature.

the scene and the subject is asked to keep track of a subset of them. Typically, an observer can accurately track up to four or five disks. Pylyshyn and Storm (1988) argue that an index is set up for each of the targets.

1.2.1.2 *Interrupt theories*

Interrupt theories, in the Green's sense (Green, 1991), can still be classified as data-driven processes in which observers detect targets by a *difference signal* arising from a discontinuity in space (Julesz, 1986) or in time (Yantis and Jonides, 1984). The discontinuity signal automatically attracts attention towards the location of 'substantial' differences, without identifying the features.

1.2.1.2.1 The texton theory. According to the glossary of computer vision terms (Haralick and Shapiro, 1991) texture is concerned with the spatial distribution of image intensities and discrete tonal features.

> 'When a small area has wide variation of discrete tonal features, the dominant property of that area is texture. There are three things crucial in this distinction:
> i) the size of the small areas; ii) the relative sizes of the discrete tonal features; iii) the number of distinguishable discrete tonal features'.

Consequently, in computer vision a *textel* (acronym for texture element) is characterized by a triplet.

Nevertheless, from the perceptual viewpoint, the textel of interest, called *texton* is characterized by the presence of elementary pattern primitives like line segments, crossings and terminators of line segments, and blobs with particular orientation, or aspect ratio, or scale and intensity (Julesz, 1981).

According to Julesz, the visual scene is pre-attentively represented as a map of differences in textons. That is, local areas that substantially differ from their neighbors in certain features are detected, even if their feature identities are not recognized. The pre-attentive system knows where a texton gradient occurs, but is unaware of what these textures are. Note that, for comparison, target detection requires at least one neighboring element to form a feature gradient.

The pre-attentive system can be operated by a feature detector with a confined local analysis, possibly implemented through the lateral inhibition between adjacent feature detectors. Feature gradients are enhanced similarly to luminance gradients, by means of center-on/center-off mechanisms (Sagi and Julesz, 1987).

An interrupt theory for spatial attention has been proposed, where it has been conjectured that any discontinuity in the texton gradient causes attention to be focused on its location. Furthermore, it is only by the attentional aperture that the textons which originated the discontinuity can be consciously known.

Any item that the pre-attentive parallel system cannot detect, requires form recognition and thus serial inspection by focal attention. According to Sagi and

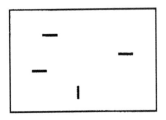 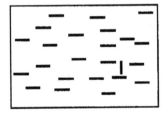

Figure 1.3 The higher the number of items, the higher the texton gradient, and the easier becomes odd element detection.

Julesz (1985), the parallel system is probably limited to the detection of feature gradients. Even such a simple task as discriminating between vertical and horizontal lines requires a serial process.

The latter operation mode leads to important consequences in performance behavior, so that in most cases, the higher the number of distractors, the closer the items lie. Moreover, the higher the texton gradient the higher is the accuracy of detection[3].

In a response time experiment (Bacon and Egeth, 1991) a *negative search function* resulted from the detection of target/distractor(s) of simple geometric forms (colored squares and oriented lines[4]).

The texton theory of pre-attentive vision is locally based: where the texton gradient overwhelms a threshold, an interrupt occurs. The distances between textons, and the aperture of focal attention are scaled by the average size of the texture elements. These three parameters make up the triplet of pre-attentive texton perception (Julesz, 1986).

Pre-attentive vision directs attentive vision to the location of interest, it is a pre-conscious process, and ignores the positional relations among the adjacent textons. Only in the conscious small aperture of focal attention are textons glued together to form shapes. Attentive vision is conscious, nevertheless it is confined to the aperture of focal attention.

1.2.1.2.2   Attentional capture. Selective attention is model driven, but when attention is captured by bottom-up data-driven stimuli, which are irrelevant to the current task, we speak of involuntary spatial attention. A new event, like the abrupt appearance of an object in the visual field, draws attention toward the *ex novo* object. Yantis and Jonides (1984) proposed that there is a mechanism which is tuned to abrupt onset and directs visual attention to the onset location. The outcome of this mechanism is the (efficient) identification of the newcomer at that location.

[3] The experimental set-up is backward masking in which the target-stimulus is followed by a masking one. The independent variable is the time between stimulus and masking onsets. Accuracy is evaluated as the percentage of correct detections.
[4] In the quoted paper, two experimental factors were considered separately: the minimum target/distractor distance and the number of distractors. By adopting the response time the results became independent of the minimum target/distractor distance!

In visual search letter recognition experiments (in which the subject knows that there is no relationship between the appearance of the target as onset and no onset[5]) the response time did not increase with set size when the target was an onset, while there was a significant positive slope when the target was a no-onset. This suggests that the abrupt onset object is processed first and that selective attention is captured at that location automatically.

Yantis and Jonides (1990) assumed that attention is implemented in the form of a priority list. The priority list contains tags that specify the current priority of unidentified objects. The magnitude of the priority tags depends on both goal-directed or top-down factors (e.g. positional expectancies or knowledge about the target attributes) and by stimulus-driven or bottom-up factors (e.g. abrupt visual onset).

Selection driven by onset stimuli is arbitrated by an attentional interrupt signal that updates the priority list when a new object description is generated. A new object description (object file for some authors) is created when a change in a stimulus attribute causes a structural change in visual scene perception, and this in turn captures attention (Yantis and Hillstrom, 1994).

The only stimulus attribute that has been shown to exhibit attentional capture is abrupt onset (Yantis, 1993), while singletons are unable to do so (Jonides and Yantis, 1988).

In the case of singletons that pop out by color or form there is no flattening in the search function slope but only a decrease depending on the saliency of the object popping out. This indicates that an object that pops out does not capture attention. The only outcome is an increased processing speed that decreases the slope, reflecting *graded attention* (Todd and Kramer, 1994).

The presence of singletons (e.g., color) known to be irrelevant interferes with attentive discrimination of another singleton (e.g. shape) which is task relevant (Theeuwes, 1992, 1994)[6]. This suggests that the bottom-up mechanism directs attention without discriminating between singletons (Pashler, 1988).

1.2.2   Attentive theories

The processing capability of the visual system is resource limited and it is not possible to deal with the huge amount of information that visual sensors gather in an exhaustive mode. For a long time attention has been conceived as a way of contending with the enormous quantity of data information with only limited processing resources.

---

[5] No-onset stimuli are presented by illuminating the stimuli in advance and camouflaging them by irrelevant line segments. The camouflaging segments are then removed, and the no-onset stimuli are revealed. On the other hand, onset stimuli (also named *abrupt onset*) appear *ex novo*.

[6] If the bottom-up mechanism is prevented by disrupting the singleton saliency (for example by increasing dissimilarities between distractors) then the irrelevant singleton does not interfere any more (Bacon and Egeth, 1994).

The capacity models of attention (Broadbent, 1958; Verghese and Pelli, 1992) assume that the limited resources can be distributed across a spatial region, that can vary in size by an order of magnitude, depending on the task. Some capacity models postulate strictly serial processing (e.g., Treisman and Gelade, 1980; Wolfe *et al.*, 1989); others postulate a parallel stage with limited resources (e.g., Duncan and Humphreys, 1989). All these models ascribe the decrease in performance, following an increase in the number of items to be processed, to an information bottleneck in the serial case, or to the need for subdivision of the attentional resources among the foveated items in the parallel case.

New models are emerging that explain the same decrease in performance more simply, as due to purely statistical reasons: the greater the number of items in the scene, the greater the likelihood of an erroneous decision (Palmer *et al.*, 1993). These noisy decision models explain attention as a contraction (under voluntary control) in the number of possible targets, without any change in the actual local processing.

Two kinds of attention are usually considered:

- *Selective attention.* It is the *model-driven* case of the search for a known target and regards how appropriate an instance selected from the scene is.
- *Spatial attention.* It is the mechanism which orients attention to a particular location in the scene. This process may be voluntary or involuntary, in the latter case it is *data driven* and involves an odd target.

1.2.2.1   *Selective attention*

Like for all resource limited systems discrimination of the peculiarities of the case at hand is critical. In most vision problems the effective implementation of a goal is strongly dependent on the capability to select the proper features for the recognition, location or interpretation processes. Once the basic properties of the target are known, different model-driven modalities can intervene to realize the visual goals. They will be discussed in context of the common experimental paradigm of visual search in the following.

1.2.2.1.1   Visual search. The paradigm of visual search was initially conceived primarily to distinguish between serial processing (one at a time, sequentially) and parallel processing (all together, simultaneously) of several distinct items. The serial processing modality assumes that the average processing time for all individual items is independent of the number of items, their position or their identity. Consequently the set-size functions, by plotting the dependence of mean RT on the size of the search set, are linear and increasing.

Moreover, visual search could be a self-terminating process in which the search stops as soon as the target is located, or it could be an exhaustive process, in which the search always extends through the entire search process, in which the search always extends through the entire search set[7]. In the former case, with a serial search, the positive set-size function is characterized by a slope equal to one half that of the target absence. This is justified by the fact that when the target is present, on average it is found half-way through the search. Instead, when a target is present and processing continues until all items in the search set are scrutinized, there should be no difference between the slopes for target presence or absence.

1.2.2.1.2 Feature integration. While singletons covering the entire visual field can be detected in parallel, attentive serial modality is necessary when target selection is based on more sophisticated analysis. Two particularly important cases are the following: i) target and distractors share features, and discriminability must be based on minor quantitative differences, e.g. target and distractors differ only by a little tilt; ii) target and distractors are distinguishable by a combination of features (in the sequel *features conjunction*), e.g. the target is a red X among distractors composed of green Xs and red Os. Both these cases require a serial search because attention must be narrowed to check one subset of items or only a single item at a time but with precision. Indeed, in the former case it is necessary to increase the signal-to-noise ratio to overcome the low discriminability. Instead, in the latter case, the need to remove the risk of *illusory conjunctions*[8] arises (Treisman and Gormican, 1988).

In Figure 1.4 a theoretical framework, suggested by Treisman to explain the role of attention in feature integration, is drawn schematically. Various components are singled out: i) a set of *feature maps*, each gathered in parallel each corresponding to a different primitive feature like a color component, a particular orientation, and so on; ii) a master *map of locations* that specifies *where* items are in the scene, but not *what* they are (not which features occupy what locations). The master map is linked to all the specialized feature maps; iii) the *attention window*: a restrained area within the master map of locations is settled so that the activity is limited to the features that are currently in the attended location (they are then bound all together), temporally excluding the others from further processing; iv) the *temporary object representation*: the feature binding forms the current item representation through a *recognition network*, the temporary object representation is compared with the object descriptions collected in the long term memory (LTM).

Activation is pooled for each feature map, within the attention window giving an average measure of the degree to which each feature is present. Thus attentive processing varies along a continuum from completely divided attention spread over the scene as a whole to sharply focused attention on one item at a time (Treisman and Gormican, 1988). Between these extremes, the attention

---

[7] The exhaustive case is experimentally found each time the target is absent from the scene.

[8] Usually in the scene there are different objects each having different features. Single feature detection is (parallel) quite effective but location is imprecise. Objects which set aside, can mix their features creating illusory conjunctions.

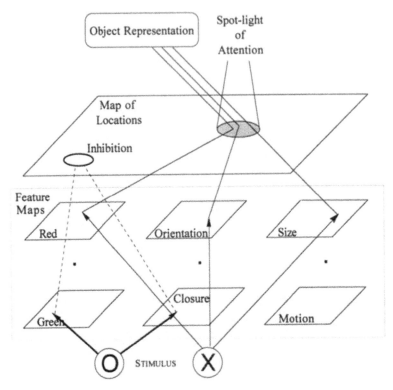

Figure 1.4 The framework proposed by feature integration theory. The red X stimulus is analysed in parallel through its primitive features in individual feature maps. The spotlight of attention acts into a map of locations, binding together the features within the focused area. This binding constitutes the current object representation. A feedback inhibition improves effectiveness by discarding locations containing different attributes.

spotlight can be set at varying intermediate sizes. The finer the grain of the scan, the more precise the location and, as a consequence, the more accurately conjoined the features present in the different maps will appear.

The size of the attention window can be adjusted to establish an appropriate signal-to-noise ratio for the target features that are currently in the attention window (Treisman, 1991).

With presentations which are too brief, during which attention may not be focused quickly enough (Treisman and Schmidt, 1982), or when different objects with similar features are too close together (Cohen and Ivry, 1989), the features of the target may be mislocated or wrongly conjoined. This is the case of illusory conjunctions which however are controversial for some authors (e.g., Green, 1991).

From the computational point of view, the feature integration theory is attractive for most artificial systems. Feature comparison is executed only within a local area and the operations are simple and can easily be performed in parallel. All the basic forms of image processing parallelism (at pixel level, at operation level, at neighboring access level, at image level) can be exploited.

1.2.2.1.3 Target/distractor similarity. Attentional strategies can be interpreted as the best allocation of limited resources to salient data (Kahneman, 1974; Burt, 1988). In this way resource assignment becomes highly competitive (Duncan and Humphreys, 1989; Rumelhart, 1970). In fact, it is necessary to discard most of the information by only selecting the stimulus factors with the highest significance for the problem at hand.

During visual search experiments, subjects are asked to detect particular target stimuli presented among irrelevant distractors. To interpret the experimental results, Duncan and Humphreys suggested four general principles: i) search efficiency varies continuously across tasks and conditions, there is no clear implication of a dichotomy between serial and parallel search modes; ii) search efficiency decreases with increasing target/distractors similarity; iii) search efficiency decreases with decreasing distractor/distractor similarity; iv) the preceding two factors interact to scale each other's effects.

Increasing target/distractor similarity produces a relatively small effect when distractor/distractor similarity is high. Decreasing distractor/distractor similarity produces little effect if target/distractors similarity is low. Alterations in search efficiency can be understood by considering both these variables together.

Similarity is fairly unambiguous when stimuli differ along simple dimensions like size or color. Proximity in such cases is easy to evaluate and is closely associated with discriminability for the human observer. For complex stimuli, such as characters differing in shape, evaluation is more complex. Such stimuli have multiple attributes, so they can be categorized in different ways. For example, a pair of characters can be similar in one respect and different in another (Beck, 1966). The relative importance of different stimulus attributes depends on the psychological context.

Following Duncan and Humphreys (1989), the human visual system is assumed to have three basic stages:

- Perceptual description produces a structured representation of the visual field at several levels of spatial scale. The whole description is derived in parallel, it is hierarchical, but does not produce immediate control over behavior. This description is phenomenologically outside awareness: it is pre-attentive.
- The selection process accesses the visual short-term memory (VSTM) on the basis of the matching between perceived descriptions and an internal template of the information required for current behavior.
- The selected information contained in the VSTM consciously controls current behavior: e.g. a structural unit may be described and picked up.

The visual representation is segmented into structural units with a hierarchical description. The parts that are within the same unit are linked together or, alternatively, parts that are used separately are described disjointly, using structural unit boundaries (Gestalt grouping). A hierarchical representation is easily obtained by repeating segmentation at different scales. Each structural unit is further sub-divided into parts by the major boundaries within it, and may include a set of properties such as the relative location, motion, color, surface texture, size and shape, and even non-visual properties such as categorization based on meaning. Similar approaches have been proposed by several computer vision researchers. In this field, three fundamental proposals are: pattern tree object representation (Burt, 1988); acyclic graph representation (Dyer, 1987); and representation through the production rules of a context-sensitive grammar (Cantoni *et al.*, 1993).

Access to VSTM is competitive and limited. Some resources exist that can be divided into varying proportions among the different structural units. Each structural unit seems to be characterized by some weight, reflecting the strength with which it competes. Two factors that determine weight selection are:

- the matching of each segment with a target template of the currently needed information. The higher the match the higher the weight. Attention is usually guided by the search for some particular piece of information which specifies the template it is referring to: in some cases this may be only one attribute of the desired information, such as its location or color; in others many combined attributes, including shapes.

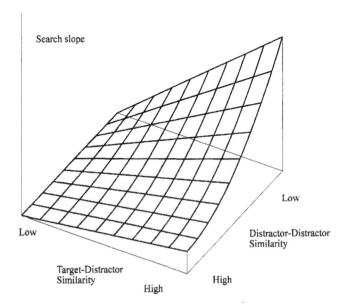

Figure 1.5 The search surface according to Duncan and Humphreys. The higher the target/distractor similarity or the lower the distractor/distractor similarity, the higher the search slope.

- weights co-vary according to grouping strength. In fact, weights tend to change according to the extent to which structural units are linked in the input segments (i.e. according to the strength of grouping between them). Accordingly, units tend to gain or lose activation together and are often jointly removed (*spreading suppression*).

The limited capacity of VSTM conditions the numbers of structural units or descriptions. When VSTM is filled, it must be flushed before new information can enter. According to some authors (Loftus, 1981) flushing is normally accompanied by refixation.

As shown in Figure 1.5 a continuum of search slopes can be found by exploiting target/distractors and distractor/distractor similarities. Increasing target/distractors similarity is harmful because during competition for VSTM access, the weight of each distractor depends on its match to target templates. Decreasing distractor/distractor similarity is also harmful because it reduces the opportunity to spread suppression of grouped distractors. In this connection, during parallel search response behavior strictly depends on how this 'whole' background is perceived (Humphreys *et al.*, 1986).

1.2.2.1.4   Perception vs decision. According to Broadbent (1958) and Treisman and Gelade (1980), visual attention can be interpreted as an *early* process which leads to the identification of the item contained in the focused area. For other authors (for example, Norman, 1968; Duncan and Humphreys, 1989) attention is a *late* process which follows perception and concerns response selection.

To compare the two hypotheses (Palmer *et al.*, 1993) analysed the set-size effect in visual search. The first hypothesis, named *perceptual coding*, conjectures that the quality of the perceptual 'code' for individual stimuli is degraded for larger set sizes. The second hypothesis named *decision integration*, holds that individual perceptual codes are unaffected by set size, which has relevance only as a noise problem in the decision phase (interpreted as a many-to-one mapping from percepts to voluntary action), thus decision is degraded with larger set size as a result of noise accumulation.

In their experiments, Palmer *et al.* controlled target/distractor similarity with a measure of accuracy based on psychophysical threshold estimation[9], that was exploitable for different stimuli and psychophysical tasks. The authors tested simple stimuli in which target and distractor differed by a single attribute such as line length, line orientation (Palmer *et al.*, 1993), brightness, or color (Palmer,

---

[9] The parameter of reference is the percentage of correct answers. It is usually set at 75%. The dependence of a feature versus set size is analysed and the feature value which corresponds to the given percentage of correct answers is identified. For example in searching for L among T the effect of shifting the two perpendicular segments in character T (always maintaining a contact point: the extreme case is an L eventually rotated character) is analysed as a function of set size, and the value giving 75% correct detection is identified. In this way, target/distractor similarity can be compared for different types of stimuli.

1994) as well as more complex stimuli (L among Ts; Palmer, 1994): a stimulus that classically leads to a serial search.

The outcome of these experiments was that the set-size effect in visual searching was *fully* accounted for by the decision model. In other words, for those stimuli, the positive search slope is explained by the increase in decision noise, that is, in the probability that incorrect choices will be made as the number of items increases.

Another important result was that endogenous cues improve detection accuracy to the extent that was expected by reducing decision noise to the relevant (that is, cued) items (Palmer *et al.*, 1993).

When the items are closely spaced the results and the theoretical provisions come unstuck; this for large set sizes might indicate an effect due to a perceptual (not attentional) change: the items lose their oneness and become micro-elements of a texture.

The authors proposed a model in which the pre-attentive stage is of unlimited capacity. The selection stage is under instructional control (voluntary attention) and not determined by immediate stimulus.

The last point stresses the idea that attentional phenomena, including so called stimulus-driven attention (e.g. attentional capture and exogenous cues), do not have sensory effects (Shiu and Pashler, 1994).

1.2.2.1.5   Activation map. Let us consider the case of a search for a red X among green Xs and red Os. This is a target defined by color and form feature conjunction, therefore, pop-out cannot occur, and the target cannot be located by a parallel process. Nevertheless, a parallel process for color detection can differentiate between green and red items supporting search efficiency.

So, to avoid wasting time and effort examining unfruitful item locations, two solutions are possible: i) the search is restricted to the locations of the red items (Wolfe *et al.*, 1989); ii) the locations of the green items are 'suppressed' (Treisman and Sato, 1990).

In visual search experiments (Wolfe *et al.*, 1989) found a wide variation in performance between subjects. This suggests that there might be some factor that varies from individual to individual that could lead to a conjunction search with evident positive slope (*serial conjunction search*) or to a conjunction search with an almost flat slope (*parallel conjunction search*). This difference between subjects is interpreted as the ability to guide the focus of attention only to the items that are most likely to contain the target.

This strategy of using the parallel processes to guide the serial process for target location, will not work for all possible search tasks, such as the search for T among Ls. In this case, both the target T and the distracting Ls are made up of a vertical and a horizontal line. A feature map sensible to orientation would 'see' the same features at all item locations. Even over a longer time span, the feature map would be unable to guide the attentional spotlight, and the resulting searches would be serial.

Feature salience plays an important role in determining whether the search for conjunctions is serial or parallel[10]. A mechanism that allows the serial stage to make use of the ability associated with the parallel maps to sub-divide items into distractors and candidate targets, has been proposed. Each parallel feature map excites all the spatial locations of candidate targets in a map of locations that supports the serial, attentional process. In the quoted example where the red Xs are to be located among green Xs and red Os, the feature map corresponding to the color red excites all the red item locations, at the same time the feature map corresponding to the X form will excite all the X item locations, so that red-X location (if any) will be doubly excited. If the spotlight of attention is directed to the point of maximum excitation, it will find the target without needing to conduct a random serial search (Cave and Wolfe, 1990).

This model implies that top-down guided searches (on the basis of target nature) can reach the parallel processes or, alternatively gate their outputs. The efficiency of this search enhances the random serial search, nevertheless a serial mechanism examines ambiguous items. If exchanges from the parallel to the serial stage are precise, attention is guided directly to the target, and conjunctive searches become independent of set size.

The degraded guiding of attention can be interpreted by assuming that the signal from the parallel to the serial processes is noisy, so that the stimulus salience affects the signal-to-noise ratio. Moreover, noise is obviously present in systems such as the 'simple' neural hardware implementing the parallel stage which must operate very quickly.

In spite of the conjunction theory, according to Cave and Wolfe (1990) there is no qualitative difference between searches for singletons and searches for conjunctions: both cases require a serial stage and the difference is due to the ability of the parallel maps to guide the attention.

For as long as the stimuli are visible the parallel processes continue to excite the activation map, so that even a small signal can emerge from noise over time (Wolfe *et al.*, 1989).

In the visual field each location holds an activation value for each feature map (e.g. one activation for color, another for orientation, etc.). Each activation value consists of a bottom-up and a top-down component. The former does not depend on the expected target and stays the same even when nothing is known about the target features. It depends on the mismatch between the value in that location and the values for the entire visual field. For a color homogeneous visual field, with a zero gradient, the bottom-up component for color activation is low everywhere. The latter component contributes to singleton search, but the most important effect is produced in conjunction searches. The top-down component is simply given by the knowledge of target features. More precisely,

---

[10] In particular, the difference in slopes between Wolfe and Treisman's experiments is accounted for by the different stimulus brightness: Treisman and Gelade (1980) presented stimuli with a tachistoscope, whereas Wolfe *et al.* used a CRT.

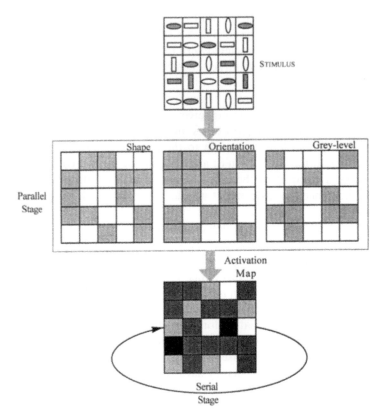

Figure 1.6 The activation map theory applied to a triple conjunction. The target is the horizontal grey rectangle (two items) and the distractors differ in shape, orientation or grey level. The parallel stage analyses the individual features; the activation map cumulates the single activities and serially guides attention towards the two most active locations.

it depends on the similarity between the value at that location and the corresponding target value. For example, if the target is vertical, then each location with a vertical element has a high activation in the orientation map (Cave and Wolfe, 1990).

Both the top-down and bottom-up components are summed together for each location of the single feature map. In turn, the activation values for each feature map are summed to produce a single overall activation map, which represents the parallel stages contribution to stimulus/target matching at each location of the visual field.

The serial stage operates over a limited area at any one time, and is composed of a collection of complex processing mechanisms. The areas are ordered by the activation level, and then sequentially analysed.

In the case of target absence the serial stage does not require every item to be processed: the analysis ends when the activations of all the unprocessed items go below a certain threshold.

On the basis of the described experimental results Treisman and Sato (1991) updated the model proposed for the feature integration theory, introducing a feedback effect between the feature maps and the location map. In fact, this top-down mechanism inhibits locations containing unlikely features, that is, they exclude those items having strong mismatch with the target features from search.

1.2.2.1.6 Features vs. surfaces. Treisman's model accounts for parallel search by detecting activity in a feature map, and conjecturing that high-level processes have access to low-level feature information. Some authors (He and Nakayama, 1992, 1994) prove that in visual search and texture segregation tasks the items which lie behind occluders are completed. In particular, they suggest that visual search operates at a level of surface representation which is beyond the level of primitive features detected by low-level filters, and that local filter activity is influenced by higher-level surface processing. Nevertheless completion is an early stage phenomenon that cannot be voluntarily overwhelmed, as shown by Kanizsa (1979).

These ideas are epitomized well in Grossberg and co-workers' model. For a complete review see Grossberg (1994). This model is supported both by neurophysiological and psychological data. The visual system is sub-divided into two streams: the boundary contour system and the feature contour system. The first delimits segments in the visual inputs by detecting primitives like edge, texture, shading, stereo and motion; the second fills in these regions with brightness, color, depth and form. A model of attentional selection was proposed (Grossberg *et al.*, 1991) to simulate visual search.

The possibility to manipulate and reorganize the segments extracted so far, on the basis of both the bottom-up and top-down mechanisms is central to this model. These operations can be recursively applied to emergent groupings allowing a multi-step search at different resolutions within the scene. Hence, the visual search is modeled in four steps: i) pre-attentive processing of the visual scene into retinotopic feature maps; ii) segmentation into boundary and filling-in of surfaces generated by multi-item grouping; a top-down priming signal from the object recognition system can influence the reorganization of segments; iii) selection of a candidate region; this selection is influenced by bottom-up saliency and top-down priming; iv) comparison of grouped features of the selected segment with the target representation. If a mismatch occurs 'attention' returns to step iii) and a new candidate segment is selected. A partial mismatch between the features of a multi-item candidate region and the target features may trigger a more vigilant search within the candidate region. The system returns to step ii) for further segmenting of the candidate region into sub-regions on the basis of a new feature dimension. The search terminates when a satisfactory match is found.

1.2.2.1.7 Multi-resolution pyramid. Selective attention exploits a trade-off between resolution and spatial extent: the same channel can be activated to attend to a large visual area with low resolution or to a small visual area with fine resolution. This trade off is needed because of the limited amount of information that attentional mechanism for high-level processing can select. For example, the particular pattern to be recognized relies on the observer's judgment and in most cases different judgments lead to different spatial resolutions.

Nakayama (1990) suggested that the visual system is composed of both a

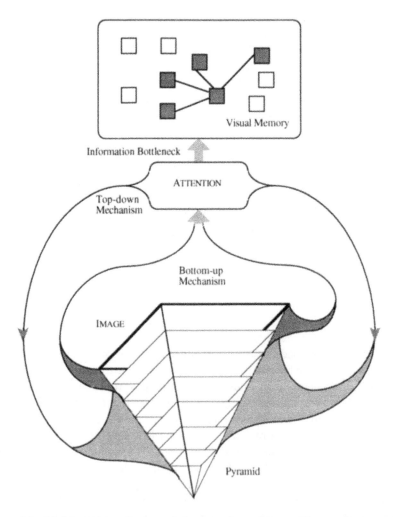

Figure 1.7 Multiresolution framework for attention guidance. The top-down mechanism guides attention towards a selected feature. The bottom-up mechanism directs attention to an odd target.

pyramidal analyser and an iconic visual memory: 'without activation {of the icons into the memory} visual perception cannot exist'. The visual field is analysed in parallel at different levels with different degrees of resolution. The memory contains a set of icons linked by associative connections. Each icon consists of a very small amount of information and because of the limited attentional bandwidth, the pyramid resolution of the segment to be compared cannot exceed icon size: hence there is a *bottleneck* between the two components represented in Figure 1.7.

In searching for a target that is defined as the odd element among distractors, two mechanisms can be used to direct attention: i) a top-down mechanism, which is exploited when the target is known; ii) a bottom-up mechanism, activated when the target is unknown and the search is for a possible odd object. These two modes of processing are known as *feature search mode* and *singleton detection mode*, respectively. Note that, when the target feature is known, both these strategies are available, however, only the latter strategy is possible for the odd target.

Koch and Ullman (1985) suggested a mechanism for directing attention by a bottom-up process. The local (lateral) inhibition within feature maps is responsible for the detection of target conspicuousness and its subsequent status of focus of attention. The activity at each location is summed across the feature maps, and attention is directed to the most stimulated location. Since the inhibitory connections between detectors in each feature map are thought to be local, it is reasonable to suppose that this process is most effective when the distractors are close together.

Summarizing, two factors determine search performance: the resolution required to make the judgment and a priori knowledge about the target. Bravo and Nakayama (1992) experimented with both these factors by utilizing stimuli containing an odd element. The authors compared different search tasks that required different resolutions. A low resolution task is used to detect or identificate the target's unique color; this can be done at low resolution. In these cases the RTs is not affected by the number of distractors in the scene (*flat search function*). Here no effect due to target knowledge resulted, because the detection of an odd target can be achieved without focusing attention.

On the contrary, if a sophisticated discrimination is required, a small aperture of attention is necessary. However, a critical aspect concerns a priori knowledge of the target's feature. In fact, this knowledge allows to direct attention, by a top-down mechanism, to the location of the highest activity for that feature. The effectiveness of this mechanism turns out to be independent of the number of distractors, as it has a characteristic feature not belonging to the target (flat search function).

Bottom-up directing of attention is needed when the unique target feature is not known. In accordance with the lateral inhibition model this is a local process. In this case, it was seen that increasing the number of distractors decreases the RTs (*negative search functions*) (Bravo and Nakayama, 1992).

Finally, in the case described by Treisman and Gelade (1980) mentioned above, in which a serial search through all items in the scene is required (discriminating between target and distractors which share some features) a *positive search function* ensues.

The multi-resolution approach has been widely investigated by the computer vision community, a survey of the suggested and implemented proposals can be found in Cantoni and Ferretti (1994).

1.2.2.2 *Spatial attention*

Visual attention orienting is usually defined in terms of foveation towards a stimulus (overt orienting). Stimulus foveation should improve the efficiency of target processing by exploiting acuity. However, it is also possible to attend to an item location covertly, without any change in eye or head position (Remington, 1980).

In psychological covert attention experiments the cueing of a spatial location is done in two ways: i) by informing the subject through a symbolic cue about the probable location of a stimulus (e.g. an arrow, displayed in the fixation point, pointing to the location, usually a box); ii) by displaying a spot of light appearing abruptly at that location. In the former case attention is oriented voluntarily and the cue is classified as *endogenous* (central cue for some authors); in the latter case attention is oriented automatically, and the cue is termed *exogenous*. Jonides (1981) proposed three characteristics to distinguish between the two types of orientation: i) the automatic orienting cannot be interrupted; ii) it does not depend on the probability of target appearance; iii) there is no interference with another, secondary, task. Controlled orienting has complementary properties. Muller and Rabbit (1989) showed that automatic and controlled orienting are based on two different mechanisms.

Most studies on spatial attention have used a cost-benefit paradigm (Posner and Boies, 1971) in which the validity of the cue is manipulated[11]. In general, RT to events occurring at the cued location (valid cue) are shorter than for neutral condition (when cue is uninformative); whereas it is longer for stimuli appearing at uncued locations (invalid cue). These effects have been classically interpreted as evidence of covert movement of attention at the cued location.

Different models have been proposed: i) the spotlight model (Posner, 1980) in which attention is conceived as a beam; ii) the zoom-lens model (Eriksen and St. James, 1986) in which the focus of attention can be differently sized, the narrower the focus the greater the benefit in processing; iii) the gradient model

[11] The cue can be valid (that is, correctly indicates the location in which the target will appear) in a greater percentage of trials, or invalid (indicating a location when actually the target will appear in another place) in a smaller percentage. A neutral condition (that corresponds to uncorrelated cues/ stimuli) is used to evaluate the benefit of validly cueing a location and the cost of having invalidly given attention to an incorrect location. This paradigm is questionable, in particular for the assumption that subjects always attend to the location that has greater probability of being correct (Kinchla, 1992).

(LaBerge and Brown, 1989) in which attention corresponds to a distribution of resources across the scene with the shape and dispersion related to the direction of attention. In the two former models the beam is moved from one location to another corresponding to the shift in attention, whereas in the gradient model the attention is not moved *per se*, but rather resource gathering emerges at the cued location.

The different performance of endogenous and exogenous cueing are summarized in the following.

1.2.2.2.1 Endogenous cues. The beam hypothesis assumes that attention crosses intermediate locations when it shifts in an analogue manner from a starting point to the final point which the subject has to attend. Some initial evidence (Shulman *et al.*, 1979) seemed to support this intuition. In fact, the experimental results are not at all compelling (Eriksen and Murphy, 1987; Yantis, 1988).

In (Rizzolatti *et al.*, 1987; Umiltà *et al.*, 1991) it is proposed that the mechanism responsible for the voluntary shifts of attention shares some of the mechanisms involved with premotor programming of saccadic eye movements. The motor program is prepared both when the saccade is subsequently executed (overt orienting) and not executed (covert orienting). Thus the only difference could be that the eye movements are inhibited in covert orienting.

Following previous works by Rosenbaum (1980), movement programming is based on two independent features: the direction and the amplitude of the movement. When an endogenous cue is presented, the motor program which specifies the direction of the eye movement and the amplitude of the saccade is prepared. In case of an incorrect cue the attention movement has to be reprogrammed. If amplitude has to be modified, an adjustment is needed when activating the already programmed set of muscles. This presumably is of small cost and could be done on line. Conversely, if modification involves the direction of movement then a complete reprogramming has to be carried out because the movement concerns a different set of muscles. This second reprogramming is time consuming. The experimental finding of a 'meridian effect' when attention crossing the vertical meridian has to be re-oriented (Rizzolatti *et al.*, 1987; Umiltà *et al.*, 1991) supports this hypothesis. Moreover, any evidence for the meridian effect and for a shift in attention is absent in the case of an exogenous cue (Umiltà *et al.*, 1991; Reuter-Lorenz and Fendrich, 1992).

It is worthwhile pointing out the relationship between voluntary and automatic orienting of attention (for example due to an abrupt onset). Yantis and Jonides (1990) demonstrated that when the location of the target is known exactly, the subject can override the attentional capture produced by abrupt onsets in other locations.

1.2.2.2.2 Exogenous cues. The abrupt appearance of a cue produces the automatic orienting of the focus of attention on the cue location. When the RT is measured, the detection of a subsequent target at that location is initially enhanced; meanwhile, about 300 ms later, the same location shows an inhibited pattern with longer RT (Posner and Cohen, 1984).

Nakayama and Mackeben (1989) used an exogenous cue in a conjunction search (for orientation and color) and, by measuring accuracy with a masking paradigm, showed that the cue produces an enhancement within a range of 50–150 ms. The range depends on the stimulus type (for instance a vernier discrimination between distracting lines is enhanced up to 300 ms after the cue).

A point of discussion concerns whether attentional selection has an effect on the processing of visual stimuli or on decision making. In fact, the reduction of RT at the cued location, does not necessarily imply enhanced processing but might reflect more 'liberal' decision making (Shaw, 1984). Shiu and Pashler (1994), by using a cost-benefit paradigm and measuring accuracy, showed that actually no enhanced processing is present at the location cued by an exogenous stimulus. The reduction in RT is fully accounted for by the decision noise reduction (Palmer *et al.*, 1993) made possible by selecting the relevant (cued) item and excluding the distractors from decision process. Thus the effect of cueing a location is to reduce the actual set size, reducing the probability of one of the distractors being mistakenly taken as a target. This noise-reduction model does not assume any changes in the perceptual efficiency of the cued objects, in particular it does not require a focus of attention where the attentional resources are concentrated.

It is worthwhile pointing out the relationship that exists between orienting attention with exogenous cues and attentional capture by abrupt onsets. Folk *et al.* (1992), by considering an exogenous cue as abrupt onset, showed that the occurrence of involuntary attentional shifts is contingent on the relationship between the properties of the cue and the properties required to locate the target in the subsequent stimulus. They crossed two types of cues and stimuli: singletons (discontinuity in spatial distribution of features) and onsets (a single item appearing abruptly, yielding to a discontinuity in time). In a first case by using pop-out for cueing a subsequent search stimulus (where the cue precedes the target stimulus and is characterized by the same feature type but with a different attribute; e.g. simplifying feature type: color; cue attribute: green among white items; target attribute in the stimulus: red among white distractors) costs (for invalid cue location) and benefits (for valid cue location) were present. In a second case, a cost-benefit effect was evident when both cue and stimulus were onsets.

Nevertheless, in the two crossed cue-target conditions, no effect of cue validity was present. Folk *et al.* (1992) suggested that observers established the type of discontinuity (whether in space or in time) to control attention by adopting an 'attentional control setting'.

One can draw an analogy between these types of attentional allocation and the interrupt implementation in a digital computer. Note that, once a stimulus triggers the activity, the event that caused the interrupt is processed involuntarily.

Yantis (1993) observed that the Folk *et al.* result indicates that the search for a uniquely colored singleton (as in the experiment described) is a top-down process that can override attentional capture by abrupt onset. This also happens for example with voluntary attention to a known spatial location (Yantis and Jonides, 1990). In fact, the interference found by Folk *et al.* cannot be considered an involuntary process because the attentional control setting based on a feature-type is a top-down process. Thus, Yantis concludes, the Folk *et al.* result only demonstrates that top-down attentional control set cannot be engaged or switched over a very small time scale between the cue stimulus and the search stimulus, so that this result is devoid of any commitment to the attentional capture theory proposed.

In this connection (Stelmach and Herdman, 1991) have shown that attention following an exogenous cue can influence the apparent timing of events: if attention is directed to a particular location, a subsequent brief flash at that location seems to occur earlier than an identical simultaneous flash placed elsewhere.

Hikosaka *et al.* (1993) have shown that this effect on timing can also induce the perceived direction of apparent motion.

When attention is moved elsewhere following an exogenous cue the processing of stimuli at the previously cued location is inhibited. This important conduct is called *inhibition of return*.

On the other hand, no inhibition occurs following an endogenous cue. Moreover, Rafal *et al.* (1989) showed that inhibition is also present just after the subject prepares an eye movement to a target location without accomplishing it. Therefore, a close link has been suggested between attention orienting to an exogenous cue and eye movement mechanisms (Posner *et al.*, 1985).

Abrams and Dobkin (1994) showed that inhibition of return has two components, one is related to eye movement inhibition (quoted above), and a second cognitive one to target detection.

In effect, they showed that the latter form of inhibition is object centered (Tipper *et al.*, 1991), that is, it follows the object when it moves to a new location.

The inhibition of return is particularly important in avoiding further exploration of locations and items that have already been inspected. The eye movements, which take a long execution time and are followed by a refractory period, require a particular effective control mechanism.

1.3 ATTENTION IN NEUROSCIENCES

Technological developments over the last decade, leading to new instruments (imagery from PET, MRI, new neuronal tracers, etc.) and consequently new and more detailed data, allowed theories, that even in the recent past could only be conjectured, to be validated and proved. As a consequence, new insights were provided into biological solutions for attentional mechanisms and in general for the primate visual system. In particular, several authors proposed and questioned whether there was a dichotomy in the visual system differentiating the two visual streams that respectively support *where* and *what* in visual processing.

1.3.1 Two visual streams

The visual system of primates is organized into a hierarchy of cortical areas as shown in Figure 1.8. The input goes from the retina to the striate cortex (V1) via the lateral geniculate nucleus of the thalamus (LGN). Within the retina, at the optic nerve connection, there are two main classes of cells: the magnocellular (M) and the parvocellular (P) ganglion cells. The projections of these two populations maintain a complete degree of independence at LGN level, whereas

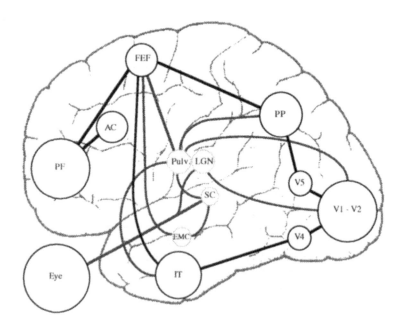

Figure 1.8 A simplified scheme of cortical areas and sub-cortical structures involved in visual perception and attention. The localization is only indicative. The structures are: AC – Anterior Cingulate (in medial frontal cortex); EMC – Eye Motor Control nuclei (in the pons); FEF – Frontal Eye Field cortex; IT – Infero Temporal cortex; LGN – Lateral Geniculate Nucleus (in the thalamus); PF – Pre-Frontal cortex; PP – Posterior-Parietal cortex; Pulv. – pulvinar nucleus (in the thalamus); V1 – striate cortex; V2, V4, V5 – extra-striate cortex; SC – Superior Colliculus (in the midbrain).

extensive intermixing[12] is present in V1 (Yoshioka *et al.*, 1994). The properties of these two ganglion cells are different, although partly overlapping: the M cells are more sensitive to low spatial and high temporal frequencies, whereas the P cells are more sensitive to high spatial and low temporal frequencies (Merigan and Maunsell, 1993).

From V1 visual information is analysed in extrastriate areas specialized for different aspects of the visual signal. Two major processing streams originate in V1 (Livingstone and Hubel, 1987). The ventral stream that originates from the P layers, involves areas V2 (thin- and inter-stripes), V4, and IT, and deals with the analysis of features such as shape, color, texture, and with object recognition. The dorsal stream, that originates from the M layers of V1, includes areas V2 (thick stripes), V3, V5 and PP, and is specialized in visual movement analysis and space representation also used (in frontal areas) for orientation in the environment and grasping.

A substantial cross talk occurs at many levels of the hierarchy between the two streams (Merigan and Maunsell, 1993) in particular between IT and PP areas (Ferrera *et al.*, 1992).

Receptive field size increases passing from V1 to IT cells. The IT cells respond maximally to complex foveal stimuli; their selectivity is relatively invariant over size, inversions, and color transformations. This suggests that they reflect object identity rather than local feature analysis (Miller *et al.*, 1991). Ungerleider and Mishkin (1982) suggested that the ventral and dorsal streams respectively maintain most properties of the P and M ganglion cells. This subdivision would have functional value: the P stream would carry over information about 'what' an object is and the M stream about 'where' it is. More recently, Milner and Goodale (1993) proposed that the *visual perception* versus *visually directed action* dichotomy accounts for neuropsychological and neurophysiological data more suitably.

Visual perception, regarding the 'what' stream, concerns the meanings and significance of objects and events and tends to be object-centered. It is connected with features and relationships between invariant representation, i.e. size, shape, color, etc.

The visually-directed action stream provides the visuo-motor mechanisms that support actions, such as manual grasping. In this case, the underlying visuo-motor transformations are viewer-centered. Moreover, given that the position and local orientation in the action space of the observer of a goal object is rarely constant, computation must take place *de novo* every time an action occurs.

The visual system deals with huge amounts of information. It is necessary to select relevant information. A primitive example is the largely automatic segmentation of figure from background; in almost all tasks a greater reduction

---

[12] Layer 2/3 is classified as parvocellular and layer 4B as magnocellular. New data show intermixing in mid layer 4C (that receives convergent M- and P-afferents from LGN, and that projects to layer 2/3-Interblob) and in layer 2/3-Blob (that receives inputs both from layers 4B, magnocellular, and 4A, parvocellular).

is required, e.g. the recognition of complex patterns carried out at maximal resolution is computationally demanding. An attentional window that can rapidly be shifted into position and changed in spatial scale makes the recognition problem manageable.

Van Essen *et al.* (1992) suggested that the architecture of the visual system is organized according to the principles of modularity and computational flexibility. The adoption of separate, specialized modules for different sub-tasks allows neuronal architecture to be optimized for various particular types of computation. The feedback connections from the higher centers allow the brain to reorganize its computational structure adaptively.

1.3.2 Attention in visual areas

The neurophysiology of attention in the macaque monkey brain has been extensively studied. Moran and Desimone (1985) recorded the activity of individual neurons of the V4 and IT areas while the animal was performing an attentional discrimination task at one spatial location in the presence of a distractor at a close location. They found that the suppression of the activity related to a distractor occurred only when the target at the attended location and the distractor close to it were simultaneously present within the neuron's receptive field. This behaviour corresponds to a receptive field which shrinks around the target, ignoring the distractor. The attentional effect seems to be mainly suppressive. In V4 no suppression in the cell's discharge was observed when only one item fell within a neuron's receptive field. In area IT, where receptive fields are much larger, the attentional effect covers a larger spatial range. Thus, going from V4 to IT, two consecutive stages for filtering unwanted information are applied, meanwhile, in areas V1 and V2, attentional suppression is almost absent.

Motter (1993) experimented an orientation discrimination task for lines. He showed an attentional effect in the V1, V2 and V4 areas. Focal attention towards a target changed (either increasing or decreasing) the response rate of these cells. These effects only occur in V1 and V2 for difficult tasks.

Chelazzi *et al.* (1993) studied the response properties of IT cells of monkeys engaged in a visual search task. The items (magazine pictures) were chosen for each cell so that one item yielded a strong response while the other elicited a poor one. The search task was organized so that the stimulus containing both good and poor items was constant and the target can be the good or the poor one. The results showed that all cells were initially activated independently of the target/distractor condition. About 200 ms later, the response to a good distractor (item with a strong response) was suppressed, whereas the response to a good target was maintained. About other 90–120 ms later the monkey made a saccade towards the target.

Schiller (1993) experimented using a visual search task with monkeys lesioned in the V4 area. Two types of stimuli were used in which all but one element were the same. For the first type of stimulus the target was a 'greater' element (e.g. having the larger size, or being the brighter element), for the second the target was a 'lesser' element (e.g. smaller size, or darker). The V4 lesion left the performance with 'greater' targets intact, but disrupted performance with lesser ones. This suggests that the deficits found after V4 lesions cannot be attributed to specific features, and that V4 contributes to the selection of physically less prominent items which in the early areas of the visual system elicit less neural activity than other nearby items.

A great deal of research has been dedicated to the relationship between attention and eye movements. In effect, after attention focusing of an object, the next step in exploring it, is usually to fixate that object. The PP, FEF areas and SC sub-cortical nucleus are especially important in eye movements for target selection. The PP area in the dorsal stream contains a spatial representation of the extra-personal environment; it projects to SC and FEF, amongst others. Duhamel *et al.* (1992) found cells in PP that show transient shifts in their receptive fields just before the animal makes a saccadic eye movement, so that stimuli which will fall within the receptive field after completion of the eye movement begin to modulate cell activity in advance. This suggests that cells in PP anticipate the retinal consequences of saccadic eye movements and update the representation of visual space to provide the locations of objects with respect to the observer. FEF and SC have parallel control of basic eye movement mechanisms: to disrupt saccade generation both these structures must be lesioned (Schiller *et al.*, 1987).

The SC is a structure where the superficial layers are visually driven and the intermediate ones are related to saccadic eye movement. The superficial layers, receiving afferences directly from the ganglion cells and widely from striate and extrastriate areas, probably constitute a point of convergence of the visual motion signals from all cortical areas, so that the presence of moving singletons in the visual field can be signaled independently of their motion direction (Davidson and Bender, 1991). Another hypothesis is that SC is a sensory-motor mapping in which the visual layers drive the saccade generating layers (Sparks, 1986).

Moreover, the SC is believed to be involved in triggering reflexive saccades and in moving visual attention to exogenous signals. Instead the FEF is a pre-motor area responsible for generating voluntary saccadic eye movements under strategic guidance (Bruce and Goldberg, 1985). It contains neurons with combined saccadic eye movement and visually-driven activity.

Using a visual search task, Schall and Hanes (1993) found that all these FEF neurons are activated initially, but neurons in their receptive field with a distractor close to the target location, are suppressed just before eye movement towards the target starts. This result is similar to that found by (Chelazzi *et al.*, 1993) for the IT area, suggesting that IT neurons drive the FEF neurons towards the purposive choice of the target. The translation into an eye movement is due to direct FEF projections towards the sub-cortical structures in the pons which are responsible for saccade generation (Segraves, 1992).

The pulvinar nucleus is a thalamic structure involved in both eye movement and attention filtering of irrelevant information (Robinson and Petersen, 1992). It projects extensively towards striate and extrastriate areas. Two subdivisions of the pulvinar seem to play the role of convergence point for both cortical visual signals and eye motor commands from SC. This allows to distinguish between visual motion in the external world and that produced by eye movements. A third subdivision of the pulvinar (not connected with V1, but with PP and the pre-motor cortex) concerns covert shifts of attention (Petersen *et al.*, 1987).

1.3.2.1   *Electrophysiology of visual search*

In the domain of electrophysiology, the effect of attention on the electrical signals registered from the human scalp has been studied. Hillyard and co-workers investigated the time course of target selection in search tasks by using event-related brain potential (ERP) techniques. Luck and Hillyard (1994a) showed that a specific component (N2pc, namely negative 200 ms, measured at the posterior contra-lateral regions) is present only in attentive search tasks (e.g. shape discrimination) *and* only when distractors are present. This suggests an attentive mechanism that filters out the distractors.

In other experiments, Luck and Hillyard (1994b) flashed a probe at target or at distractor location with a delay corresponding to N2pc after the search stimulus onset. They showed that i) no enhancement in processing (indicated by component P1, positive 100 ms) is present at the target location, whereas suppression is present at a distractor location but only when there is a concomitant target: target selection suppresses the distractors; ii) suppression is present only in attentive search tasks (e.g. search for shape) but it is absent in simple feature search (tasks that are supported by pop-out detection); iii) a second component (N1, negative 100 ms) is present, in both attentive and feature tasks and only at target location. Mangun *et al.* (1993) localized P1 in the ventral pre-striate cortex and N1 in occipito-parietal cortex, proving that these components cannot occur in the striate cortex. On the other hand, the N2pc component is localized in the occipital areas, presumably in the striate cortex (Luck and Hillyard, 1994).

Luck and Hillyard (cit. in Luck, 1994) proposed a model in which a first parallel stage (in V1 and extra-striate areas) is followed by a localization (through cooperation between IT and PP areas) of the likely target. This allows a signal to be sent back (via pulvinar) to the first stage. This signal then drives the flow of information allowing suppression of the distractors. The suppression allows the selected item to be identified in absence of interfering information. This is a scheme for selection-and-filtering attention (see LaBerge and Brown, 1989).

1.3.2.2 *Neuropsychology of attention*

In the domain of neuropsychology the effect of brain injuries on the disruption of attentional behavior has been studied. The Posner and Mesulam models are briefly described in the following.

Posner and Petersen (1990) suggested that three areas are involved when attending to a location: i) the posterior parietal cortex, ii) the superior colliculus, iii) the lateral pulvinar nucleus of the thalamus. These structures are involved in three different aspects of attention: i) the ability to disengage attention from a previously inspected location, ii) the ability to shift attention to a new location, and iii) the ability to engage attention on a target. The parietal cortex first disengages attention from the location of its present focus, then the superior colliculus moves the focus of attention to the new target location, and finally the pulvinar reads out data from this location. Posner called this multi-component system the *posterior attentional system* to distinguish it from an *anterior attentional system* concerned with strategic control of behavior.

A network model of the brain areas involved in attention has been proposed by (Mesulam, 1981, 1990). He has particularly addressed the neglect syndrome, that is, the inability to direct attention and to respond to stimuli presented in the left visual field in the absence of muscle weakness or primary areas damages. In monkeys it is manifested in both visual hemifields, contralateral to the lesion; in man (for right-handed subjects) after a posterior parietal lesion in the right hemisphere. In his model, directed attention is organized in a distributed network that contains three cortical components corresponding to three different aspects of the syndrome: i) perceptual (following PP damage); ii) motor (FEF damage) and iii) limbic (AC damage). These areas are connected together, each providing a different coordinate system for mapping the environment. The PP component provides a sensory representation of extrapersonal space, the FEF provides a map for the distribution of orienting and exploratory eye movements, and AC provides a map for assigning value (based on the importance of stimuli behaviour from the limbic system) to spatial coordinates given by the other two components. The SC is more closely affiliated with the motor component of attention. The sub-cortical nuclei (pulvinar for visual and prefrontal areas, striatum for PP and FEF) and feedback cortico/cortical connections provide processing synchronization between areas, supporting the mechanism for binding perception into action. Lesion in any of these areas and nuclei, or to their interconnections, can bring about the neglect syndrome.

1.3.2.3 *The dynamic routing circuit model*

Olshausen *et al.* (1993) proposed a neuro-biologically plausible model of selective attention. Their model is conceived as a multi-component system in which information routes from any restricted portion of the visual field to a high-level center for pattern recognition. The model is based on a mechanism which shifts and rescales the representation of an object flowing from its retinal reference frame towards an object-centered reference frame. The major constraint posed is that loss of detailed spatial relationship should occur within the window of attention.

There are five levels of neural arrays (with each node representing a feature vector) with feed-forward connections. The inputs of a node come from a limited number of nodes in the preceding level. These connections are gathered by control units to provide shifting of information from one level to the next. The preservation of spatial relationships is ensured by a winner-take-all network between control units connected so that the ones corresponding to a common translation or scale, reinforce each other (with excitatory weights), while control units that are not part of the same transformation inhibit each other (with inhibitory weights).

The scaling effect is achieved by choosing the dimension of the attention window. When the window is at its smallest size (same resolution as the input level), the connections are set so as to establish a one-to-one correspondence between nodes in the output and the attended nodes in the input. When the window is larger, the connections must be set so that multiple inputs converge onto a single output node, resulting in a lower-resolution representation of the contents.

The model is set to direct attention to the most salient object included in the scene. The brightest blob present in a low-pass-filtered version of the scene is detected by the control units. Then the window of attention is shifted to that position. The high-resolution content is gathered in the window, and the recognition phase is performed by memory matching. The current location is then inhibited, in order to prevent a direct return, after which attention is shifted to the next location.

The authors proposed an analogy between, on the one hand, the activity of the control units with that of the pulvinar and, on the other, the activity around the saliency map with the SC and PP operations. Five multi-resolution levels of a pyramidal paradigm, correspond to the sequence of striate (V1-infragranular and V1-supragranular layers), extrastriate (V2, V4) and IT areas respectively.

1.4 THE ATTENTIONAL EXECUTIVE SYSTEM

The pre-frontal cortical areas are involved in programming high-order behavior. A basic requirement for their involvement is the degree of centralization needed for the current action: highly automatized operations are executed without central control, whereas new actions, or actions executed in new environments, or actions effected with new mappings of perception into response, need a central system that 'modulates' the execution of pre-defined action schemes (Schneider *et al.*, 1994).

The frontal cortex is organized into a hierarchy of cortical areas in which pre-motor and motor areas are specialized in planning responses (Di Pellegrino and Wise, 1993) whereas pre-frontal areas are specialized as the working memory of sensory and temporal information (Quintana and Fuster, 1992). Overall, the brain can be represented as an ascending hierarchy of sensory areas (responding to elementary features of the stimuli at the lowest level, and to symbolic and extra-personal spatial coordinate representations at the higher levels) and a descending hierarchy of motor areas that translate perception into action: pre-frontal cortex containing broad schemes of sequentially organized actions (e.g. syntactic and logical statements in spoken language), pre-motor and motor cortices representing more specific actions that are discrete in terms of somatotopy (e.g., morphemes and phonemes) and movements and their trajectories (Fuster, 1989).

Some theories have been proposed for the pre-frontal cortex. Norman and Shallice (1980) postulated that at the highest level, action is represented by schemes such as scripts (Schank and Abelson, 1977). They are activated in connection with environmental conditions and temporally guided execution of sub-routines implemented in specialized areas. A *supervisory attentional system* (SAS) is needed to deal with the intentions and decision making of the subject, with new environmental conditions and to allow learning from errors. SAS operates by modulating the schemes quoted above.

Schallice (1988) localizes this system in the pre-frontal cortex on the basis of neuropsychological data.

The *anterior attentional system* (AAS) (localized in the anterior cingulate and basal ganglia) described by Posner and Petersen (1990) is concerned with the recruitment and control of the posterior areas of the brain that are responsible for sensory processing. These goals are often termed *executive function* suggesting that the executive AAS is informed of the processes taking place within the organization of specialized areas (phenomenologically, this would correspond to awareness) and can achieve some control over the system.

Some experimental situations are used in psychology to inquire into the limits of dealing with two tasks at a time (PRP[13]) or the inability to suppress useless information (e.g., the Stroop effect[14]). These tasks are supposed to involve an 'executive function'. It is often assumed that limited resources are available to

---

[13] The experiment often consists of two two-choice tasks presented in rapid succession (e.g. the first requiring response to a tone, the second discrimination of visual stimuli) and show a consistent slowing (PRP: psychological refractory period) in the second response (Duncan, 1980). Recent results seem to indicate a model in which sensory processing is unaffected by the dual task and the effect is due to queuing of response selection stages (Fagot and Pashler, 1992; Pashler, 1994).

[14] The stimuli consist of color words printed in colored inks; the subject must name the color of the ink. When ink and word are incompatible (e.g., RED printed in green ink), large interference becomes evident: RT slows down (Stroop, 1935). PET studies revealed strong activation of the anterior cingulate (Pardo *et al.*, 1990; Bench *et al.*, 1993).

the response system, similarly to what has been said above regarding the attentional visual system, but some results seem to indicate that the two domains are independent (Pashler, 1991).

REFERENCES

AA.VV. (1994) Dialogue. *CVGIM: Image Understanding* **60**, 65–118.

Abrams, R.A. and Dobkin, R.S. (1994) Inhibition of return: effects of attentional cuing on eye movement latencies. *Journal of Experimental Psychology: Human Perception and Performance* **20**, 467–477.

Bacon, W.F. and Egeth, H.E. (1991) Local processes in preattentive feature detection. *Journal of Experimental Psychology: Human Perception and Performance* **17**, 77–90.

Bacon, W.F. and Egeth, H.E. (1994) Overriding stimulus-driven attentional capture. *Perception and Psychophysics* **55**, 485–496.

Barlow, H.B. (1972) Single units and sensation: a neuron doctrine for perceptual psychology. *Perception* **1**, 371–394.

Beck, J. (1966) Effect of orientation and of shape similarity on perceptual grouping. *Perception and Psychophysics* **1**, 300–302.

Bench, C.J., Frith, C.D., Grasby, P.M., Friston, K.J., Paulescu, E., Franckowiak, R.S.J. and Dolan, R.J. (1993) Investigations of the functional anatomy of attention using the Stroop test. *Neuropsychologia* **31**, 907–922.

Bravo, M.J. and Nakayama, K. (1992) The role of attention in different visual-search tasks. *Perception and Psychophysics* **51**, 465–472.

Broadbent, D.E. (1958) *Perception and communication.* Pergamon Press, London, UK.

Bruce, C.J. and Goldberg, M.E. (1985) Primate frontal eye fields: I. Single neurons discharging before saccades. *Journal of Neurophysiology* **53**, 603–635.

Burt, P.J. (1988) Attention mechanisms for vision in a dynamic world. *Proc. 11th Int. Conf. on Pattern Recognition*, 977–987, Rome, Italy.

Burt, P.J., Anderson, C.H., Sinniger, J.O. and van der Wal, G. (1986) A pipeline pyramid machine. In: *Pyramidal systems for computer vision* (eds V. Cantoni and S. Levialdi), 133–152, Springer Verlag, Berlin, Germany.

Cantoni, V. and Ferretti, M. (1994) *Pyramidal architectures for computer vision.* Plenum Press, New York.

Cantoni, V., Cinque, L., Guerra, C., Levialdi, S. and Lombardi, L. (1993) Describing object by a multi-resolution syntactic approach. In: *Parallel image analysis* (eds A. Nakamura, M. Nivat, A. Saudi, P.S.P. Wang and K. Inoue), 54–68. Springer Verlag, Heidelberg, Germany.

Cave, K.R. and Wolfe, J.M. (1990) Modeling the role of parallel processing in visual search. *Cognitive Psychology* **22**, 225–271.

Chelazzi, L., Miller, E.K., Duncan, J. and Desimone, R. (1993) A neural basis for visual search in inferior temporal cortex. *Nature* **363**, 345–347.

Cohen, A. and Ivry, R. (1989) Illusory conjunctions inside and outside the focus of attention. *Journal of Experimental Psychology: Human Perception and Performance* **15**, 650–663.

Damasio, A.R. (1989) The brain binds entities and events by multiregional activation from convergence zones. *Neural Computation* **1**, 121–129.

Davidson, R.M. and Bender, D.B. (1991) Selectivity for relative motion in the monkey superior colliculus. *Journal of Neurophysiology* **65**, 1115–1133.

Di Pellegrino, G. and Wise, S.P. (1993) Visuospatial versus visuomotor activity in the premotor and prefrontal cortex of a primate. *Journal of Neuroscience* **13**, 1227–1243.

Duhamel, J-R., Colby, C.L. and Goldberg, M.E. (1992) The updating of the representation of visual space in parietal cortex by intended eye movements. *Science* **255**, 90–92.

Duncan, J. (1980) The locus of interference in the perception of simultaneous stimuli. *Psychological Review* **87**, 272–300.

Duncan, J. and Humphreys, G.W. (1989) Visual search and stimulus similarity. *Psychological Review* **96**, 433–458.

Dyer, R. (1987) Multiscale image understanding. In: *Parallel computer vision* (ed. L. Uhr), 171–213. Academic Press, Orlando, Florida.

Engel, A.K., Konig, P., Kreiter, A.K., Schillen, T.B. and Singer, W. (1992) Temporal coding in the visual cortex: new vistas on integration in the nervous system. *Trends in Neuroscience* **15**, 218–226.

Enns, J.T. and Rensink, R.A. (1991) Preattentive recovery of three-dimensional orientation from line drawings. *Psychological Review* **98**, 335–351.

Eriksen, C.W. and Murphy, T.D. (1987) Movement of attentional focus across the visual field: a critical look at the evidence. *Perception and Psychophysics* **42**, 299–305.

Eriksen, C.W. and St James, J.D. (1986) Visual attention within and around the field of focal attention: a zoom lens model. *Perception and Psychophysics* **40**, 225–240.

Fagot, C. and Pashler, H. (1992) Making two responses to a single object: implications for the central attentional bottleneck. *Journal of Experimental Psychology: Human Perception and Performance* **18**, 1058–1079.

Ferrera, V.P., Nealey, T.A. and Maunsell, J.H.R. (1992) Mixed parvocellular and magnocellular geniculate signals in visual area V4. *Nature* **358**, 756–758.

Findlay, J.M. (1982) Global processing for saccadic eye movements. *Vision Research* **22**, 1033–1045.

Folk, C.L., Remington, R.W. and Johnston, J.C. (1992) Involuntary covert orienting is contingent on attentional control settings. *Journal of Experimental Psychology: Human Perception and Performance* **18**, 1030–1044.

Fuster, J. (1989) *The prefrontal cortex: anatomy, physiology, and neuropsychology of the frontal lobe.* Raven, New York.

Green, M. (1991) Visual search, visual streams, and visual architectures. *Perception and Psychophysics* **50**, 388–403.

Grossberg, S. (1994) 3-D vision and figure-ground separation by visual cortex. *Perception and Psychophysics* **55**, 48–120.

Grossberg, S., Mingolla, E. and Ross, W.D. (1991) A neural theory of attentive visual search: interactions of boundary, surface, spatial, and object representations. *Psychological Review* **101**, 470–489.

Haralick, R.M. and Shapiro, L.G. (1991) Glossary of computer vision terms. *Pattern Recognition* **24**, 69–93.

He, Z.J. and Nakayama, K. (1992) Surfaces versus features in visual search. *Nature* **359**, 231–233.

He, Z.J. and Nakayama, K. (1994) Perceiving textures: beyond filtering. *Vision Research* **34**, 151–162.

Hikosaka, O., Miyauchi, S. and Shimojo, S. (1993) Focal visual attention produces illusory temporal order and motion sensation. *Vision Research* **33**, 1219–1240.

Hubel, D.H. and Wiesel, T.N. (1962) Receptive fields, binocular interaction and functional architecture in the cat's visual cortex. *Journal of Physiology* **160**, 106–154.

Humphreys, G.W., Quinlan, P.T. and Riddoch, M.J. (1989) Grouping processes in visual search: effect with single- and combined-feature targets. *Journal of Experimental Psychology: General* **118**, 258–279.

Jonides, J. (1981) Voluntary versus automatic control over the mind's eye. In: *Attention and performance IX* (eds J. Long and A. Baddeley). Erlbaum, Hillsdale, New Jersey.

Jonides, J. and Yantis, S. (1988) Uniqueness of abrupt visual onsets in capturing attention. *Perception and Psychophysics* **43**, 346–354.

Julesz, B. (1981) Textons, the elements of texture perception, and their interactions. *Nature* **290**, 91–97.

Julesz, B. (1986) Texton gradients: the texton theory revisited. *Biological Cybernetics* **54**, 245–251.

Kahneman, D. (1973) *Attention and effort*. Prentice-Hall, Englewood Cliffs, NJ.

Kanizsa, G. (1979) *Organization in vision: essays in gestalt perception*. Prager, New York.

Kinchla, R.A. (1992) Attention. *Annual Review of Psychology* **43**, 711–742.

Kleffner, D.A. and Ramachandran, V.S. (1992) On the perception of shape from shading. *Perception and Psychophysics* **52**, 18–36.

Koch, C. and Ullman, S. (1985) Shifts in selective visual attention: towards the underlying neural circuitry. *Human Neurobiology* **4**, 219–227.

LaBerge, D. and Brown, V. (1989) Theory of attentional operations in shape identification. *Psychological Review* **96**, 101–124.

Livingstone, M.S. and Hubel, D.H. (1987) Psychophysical evidence for separate channels for the perception of form, color, movement and depth. *Journal of Neuroscience* **7**, 3416–3468.

Loftus, G.R. (1981) Tachistoscopic simulations of eye fixations on pictures. *Journal of Experimental Psychology: Human Perception and Performance* **7**, 369–376.

Luck, S.J. and Hillyard, S.A. (1994) Spatial filtering during visual search: evidence from human neurophysiology. *Journal of Experimental Psychology: Human Perception and Performance* **20**, 1000–1014.

Luck, S.J. (1994) Cognitive and neural mechanisms of visual search. *Current Opinion in Neurobiology* **4**, 183–188.

Mangun, G.R., Hillyard, S.A. and Luck, S.J. (1993) Electrocortical substrates of visual selective attention. In: *Attention and performance XIV* (eds D. Meyer and S. Kornblum), 219–243. MIT Press, Cambridge, MA.

Marr, D. (1982) *Vision*. Freeman, New York, NY.

McLeod, P., Driver, J., Dienes, Z. and Crisp, J. (1991) Filtering by movement in visual search. *Journal of Experimental Psychology: Human Perception and Performance* **17**, 55–64.

Merigan, W.H. and Maunsell, J.H.R. (1993) How parallel are the primate visual pathways? *Annual Review of Neuroscience* **16**, 369–402.

Mesulam, M-M. (1981) A cortical network for directed attention and unilateral neglect. *Annual Neurology* **10**, 309–325.

Mesulam, M-M. (1990) Large-scale neurocognitive networks and distributed processing for attention, language, and memory. *Annual Neurology* **28**, 597–613.

Miller, E.K., Li, L. and Desimone, R. (1991) A neural mechanism for working and recognition memory in inferior temporal cortex. *Science* **254**, 1377–1379.

Milner, A.D. and Goodale, M.A. (1993) Visual pathways to perception and action. In: *The visually responsive neuron: from basic neurophysiology to behavior* (eds T.P. Hicks, S. Molotchnikoff and T. Ono), Progress in Brain Research 95, 317–338. Elsevier, Amsterdam, The Netherlands.

Moran, J. and Desimone, R. (1985) Selective attention gates visual processing in the extrastriate cortex. *Science* **229**, 782–784.

Motter, B.C. (1993) Focal attention produces spatially selective processing in visual cortical areas V1, V2, and V4 in the presence of competing stimuli. *Journal of Neurophysiology* **70**, 909–919.

Muller, H.J. and Rabbit, P.M.A. (1989) Reflexive and voluntary orienting of visual attention: time course of activation and resistance to interruption. *Journal of Experimental Psychology: Human Perception and Performance* **15**, 315–330.

Nakayama, K. (1990) The iconic bottleneck and the tenuous link between early visual processing and perception. In: *Vision: coding and efficiency* (ed. C. Blakemore), 411–422. Cambridge University Press, Cambridge, UK.

Nakayama, K. and Mackeben, M. (1989) Sustained and transient components of focal visual attention. *Vision Research* **29**, 1631–1647.

Nakayama, K. and Silverman, G.H. (1986) Serial and parallel processing of visual feature conjunctions. *Nature* **320**, 264–265.

Neisser, U. (1967) *Cognitive psychology*. Appleton-Century-Crofts, New York.

Norman, D.A. (1968) Toward a theory of memory and attention. *Psychological Review* **75**, 522–536.

Norman, D.A. and Shallice, T. (1980) Attention to action: willed and automatic control of behavior. Center for Human Information Processing, Technical Report no. 99.

Olshausen, B.A., Anderson, C.H. and Van Essen, D.C. (1993) A neurobiological model of visual attention and invariant pattern recognition based on dynamic routing of information. *Journal of Neuroscience* **13**, 4700–4719.

Palmer, J. (1994) Set-size effects in visual search: the effect of attention is independent of the stimulus for simple tasks. *Vision Research* **34**, 1703–1721.

Palmer, J., Ames, C.T. and Lindsey, D.T. (1993) Measuring the effect of attention on simple visual search. *Journal of Experimental Psychology: Human Perception and Performance* **19**, 108–130.

Pardo, J.V., Pardo, P.J., Janer, K.W. and Raichle, M.E. (1990) The anterior cingulate cortex mediates processing selection in the Stroop attentional conflict paradigm. *Proceedings of the National Academy of Sciences USA* 87, 256–259.

Pashler, H. (1988) Cross-dimensional interaction and texture segregation. *Perception and Psychophysics* **43**, 307–318.

Pashler, H. (1991) Shifting visual attention and selecting motor responses: distinct attentional mechanisms. *Journal of Experimental Psychology: Human Perception and Performance* **17**, 1023–1040.

Pashler, H. (1994) Graded capacity-sharing in dual-task interference? *Journal of Experimental Psychology: Human Perception and Performance* **20**, 330–342.

Petersen, S.E., Robinson, D.L. and Morris, J.D. (1987) Contributions of the pulvinar to visual spatial attention. *Neuropsychologia* **25**, 97–105.

Posner, M.I. (1980) Orienting of attention. *Quarterly Journal of Experimental Psychology* **32**, 3–25.

Posner, M.I. and Boies, S.J. (1971) Components of attention. *Psychological Review* **78**, 391–408.

Posner, M.I. and Cohen, Y. (1984) Components of visual orienting. In: *Attention and performance X* (eds H. Bouma and D. Bouwhuis), 531–556, Erlbaum, Hillsdale, New Jersey.

Posner, M.I. and Petersen, S.E. (1990) The attention system of the human brain. *Annual Review of Neuroscience* **11**, 25–42.

Posner, M.I., Rafal, R.D., Choate, L.S. and Vaughan, J. (1985) Inhibition of return: neural basis and function. *Cognitive Neuropsychology* **2**, 211–228.

Pylyshyn, Z. (1989) The role of location indexes in spatial perception: a sketch of the FINST spatial-index model. *Cognition* **32**, 65–97.

Pylyshyn, Z. and Storm, R. (1988) Tracking multiple independent targets: Evidence for a parallel tracking mechanism. *Spatial Vision* **3**, 1–19.

Quintana, J. and Fuster, J. (1992) Mnemonic and predictive functions of cortical neurons in a memory task. *NeuroReport* **3**, 721–724.

Rafal, R.D., Calabresi, P.A., Brennan, C.W. and Sciolto, T.K. (1989) Saccade preparation inhibits reorienting to recently attended locations. *Journal of Experimental Psychology: Human Perception and Performance* **15**, 673–685.

Remington, R.W. (1980) Attention and saccadic eye movements. *Journal of Experimental Psychology: Human Perception and Performance* **6**, 726–744.

Reuter-Lorenz, P.A. and Fendrich, R. (1992) Oculomotor readiness and covert orienting: differences between central and peripheral precues. *Perception and Psychophysics* **52**, 336–344.

Rizzolatti, G., Riggio, L., Dascola, I. and Umiltà, C. (1987) Reorienting attention across the horizontal and vertical meridians: evidence in favor of a premotor theory of attention. *Neuropsicologia* **25**, 31–40.

Robinson, D.L. and Petersen, S.E. (1992) The pulvinar and visual salience. *Trends in Neuroscience* **15**, 127–132.

Rosenbaum, D.A. (1980) Human movement initiation: specification of arm, direction, and extent. *Journal of Experimental Psychology: General* **109**, 444–474.

Rumelhart, D.E. (1970) A multicomponent theory of the perception of briefly exposed visual displays. *Journal of Mathematical Psychology* **7**, 191–218.

Sagi, D. and Julesz, B. (1985) Detection versus discrimination of visual orientation. *Perception* **14**, 619–628.

Sagi, D. and Julesz, B. (1987) Short-range limitation on detection of feature differences. *Spatial Vision* **2**, 39–49.

Schall, J.D. and Hanes, D.P. (1993) Neural basis of saccade target selection in frontal eye field during visual search. *Nature* **366**, 467–469.

Schallice, T. (1988) *From neuropsychology to mental structure.* Cambridge University Press, Cambridge, UK.

Schank, R.C. and Abelson, R. (1977) *Scripts, plans, goals and understanding.* Erlbaum, Hillsdale, New Jersey.

Schiller, P.H. (1993) The effects of V4 and middle temporal (MT) area lesions on visual performance in the rhesus monkey. *Visual Neuroscience* **10**, 717–746.

Schiller, P.H., Sandell, J.H. and Maunsell, J.H.R. (1987) The effect of frontal eye field and superior colliculus lesions on saccadic latencies in the rhesus monkey. *Journal of Neurophysiology* **57**, 1033–1049.

Schneider, W., Pimm-Smith, M. and Worden, M. (1994) Neurobiology of attention and automaticity. *Current Opinion in Neurobiology* **4**, 177–182.

Segraves, M.A. (1992) Activity of monkey frontal eye field neurons projecting to oculomotor regions of the pons. *Journal of Neurophysiology* **68**, 1967–1985.

Shaw, M.L. (1984) Division of attention among spatial locations: a fundamental difference between detection of letters and detection of luminance increments, in: *Attention and performance X* (Eds. H. Bouma and D. Bouwhuis), 106–121. Erlbaum, Hillsdale, New Jersey.

Shiu, L. and Pashler, H. (1994) Negligible effect of spatial precuing on identification of single digits. *Journal of Experimental Psychology: Human Perception and Performance* **20**, 1037–1054.

Shulman, G.L., Remington, R.W. and McLean, J.P. (1979) Moving attention through visual space. *Journal of Experimental Psychology: Human Perception and Performance* **5**, 522–526.

Sparks, D.L. (1986) Translation of sensory signals into commands for control of saccadic eye movements: role of primate superior colliculus. *Physiological Review* **66**, 118–171.

Stelmach, L.B. and Herdman, C.M. (1991) Directed attention and perception of temporal order. *Journal of Experimental Psychology: Human Perception and Performance* **17**, 539–550.

Stroop, J.R. (1935) Studies of interference in serial verbal reactions. *Journal of Experimental Psychology* **18**, 643–662.

Theeuwes, J. (1992) Perceptual selectivity for color and form. *Perception and Psychophysics* **51**, 599–606.

Theeuwes, J. (1994) Stimulus-driven capture and attentional set: selective search for color and visual abrupt onsets. *Journal of Experimental Psychology: Human Perception and Performance* **20**, 799–806.

Tipper, S.P., Driver, J. and Weaver, B. (1991) Short report: object-centered inhibition of return of visual attention. *Quarterly Journal of Experimental Psychology* **43A**, 289–298.

Todd, S. and Kramer, A.F. (1994) Attentional misguidance in visual search. *Perception and Psychophysics* **56**, 198–210.

Treisman, A. (1991) Search, similarity, and integration of features between and within dimensions. *Journal of Experimental Psychology: Human Perception and Performance* **17**, 652–676.

Treisman, A. and Gelade, G. (1980) A feature-integration theory of attention. *Cognitive Psychology* **12**, 97–136.

Treisman, A. and Gormican, S. (1988) Feature analysis in early vision: evidence from search asymmetries. *Psychological Review* **95**, 15–48.

Treisman, A. and Sato, S. (1990) Conjunction search revisited. *Journal of Experimental Psychology: Human Perception and Performance* **16**, 459–478.

Treisman, A. and Schmid, N. (1982) Illusory conjunctions in the perception of objects. *Cognitive Psychology* **14**, 107–141.

Trick, L.M. and Pylyshyn, Z. (1983) What enumeration studies can show us about spatial attention: evidence from limited capacity preattentive processing. *Journal of Experimental Psychology: Human Perception and Performance* **19**, 331–351.

Tsotsos, J.K. (1994) There is no one way to look at vision. *CVGIM: Image Understanding* **60**, 95–97.

Uhr, L. (1987) Highly parallel, hierarchical, recognition cone perceptual structures. In: *Parallel computer vision* (ed. L. Uhr), 249–287, Academic Press, London, UK.

Ullman, S. (1984) Visual routines. *Cognition* **18**, 97–159.

Umiltà, C., Riggio, L., Dascola, I. and Rizzolatti, G. (1991) Differential effects of central and peripheral cues on the reorienting of spatial attention. *European Journal of Cognitive Psychology* **3**, 247–267.

Ungerleider, L.G. and Mishkin, M. (1982) Two cortical visual systems. In: *Analysis of visual behavior* (eds D.J. Ingle, M.A. Goodale and R.J.W. Mansfield), 549–585, MIT Press, Cambridge, Massachusetts.

Van Essen, D.C., Anderson, C.H. and Felleman, D.J. (1992) Information processing in the primate visual system: an integrated systems perspective. *Science* **255**, 419–423.

Verghese, P. and Pelli, D.G. (1992) The information capacity of visual attention. *Vision Research* **32**, 983–995.

Wolfe, J.M., Cave, K.R. and Franzel, S.L. (1989) Guided search: an alternative to the feature integration model for visual search. *Journal of Experimental Psychology: Human Perception and Performance* **15**, 419–433.

Yantis, S. (1988) On analog movements of visual attention. *Perception and Psychophysics* **43**, 203–206.

Yantis, S. (1993) Stimulus-driven attentional capture and attentional control settings. *Journal of Experimental Psychology: Human Perception and Performance* **19**, 676–681.

Yantis, S. and Hillstrom, A.P. (1994) Stimulus-driven attentional capture: evidence from equiluminant visual objects. *Journal of Experimental Psychology: Human Perception and Performance* **20**, 95–107.

Yantis, S. and Jonides, J. (1984) Abrupt visual onsets and selective attention: evidence from visual search. *Journal of Experimental Psychology: Human Perception and Performance* **10**, 601–621.

Yantis, S. and Jonides, J. (1990) Abrupt visual onsets and selective attention: voluntary versus automatic allocation. *Journal of Experimental Psychology: Human Perception and Performance* **16,** 121–134.

Yarbus, L. (1967) *Eye movements and vision.* Plenum Press, New York.

Yoshioka, T., Levitt, J.B. and Lund, J.S. (1994) Independence and merger of thalamocortical channels within macaque monkey primary visual cortex: anatomy of interlaminar projections. *Visual Neuroscience* **11,** 467–489.

2

Attentional Mechanisms in Computer Vision

2.1 INTRODUCTION

Biological vision is foveated, highly goal oriented and task dependent. This observation, which is rather clear if we trace the behavior of practically every vertebrate, is now being taken seriously into consideration by the computer vision community. This is evident from recent work on active vision systems and heads (Clark and Ferrier, 1988; Brunnstrome et al., 1992; Crowley, 1991; Rimey and Brown, 1992) and general active vision concepts and algorithms (Aloimonos et al., 1987; Bajcsy, 1988; Aboot and Ahuja, 1988; Ballard, 1990; Culhane and Tsotsos, 1992). One of the fundamental features of active vision is the use of space-variant vision and sensors (Yeshurun and Schwartz, 1989; Tistarelli and Sandini, 1990; Rojer and Schwartz, 1990), that allows, in the case of the log-polar representation, data reduction as well as a certain degree of size and rotation invariance.

The use of such sensors require efficient mechanisms for gaze control, that are, in turn, directed by attentional algorithms. Using psychophysical terms, these algorithms are either *overt*, analysing in detail the central foveated area, or *covert*, analysing various regions within the field of view that are not necessarily in the central foveated area.

Like many other issues in computational vision, the attention problem seems to be trapped in the typical top-down bottom-up cycle, as well as in the global-local cycle: global processes are necessarily based on local features and processes, whose crucial parameters, in turn, depend on global estimates.

ARTIFICIAL VISION
ISBN 0-12-444816-X
Copyright © 1997 Academic Press Ltd
All rights of reproduction in any form reserved

This is the case for recognition tasks, where, for example, thresholds and size tuning of local feature detectors are optimally determined by the model of the object the system expects. In curve and edge detection, local discontinuities are classified as signals or as noise according to global matching based on these very local estimates (Zucker *et al.*, 1989).

Similarly, attention is undoubtedly a concurrent top-down *and* bottom-up process: computational resources are assigned to regions of interest. But detection of regions of interest is both *context dependent* (top down), since the system is task oriented, and *context free* (bottom up), since one of the most important aspects of such a system is detection of unexpected signals. Thus, attention must be based on highly coupled low-level and high-level processes. While we do not offer a solution to this fundamental problem, we describe in this review a number of methodologies that begin this cycle with a low-level attentional mechanism.

Visual processes, in general, and attentional mechanisms, in particular, seem effortless for humans. This introspection, however, is misleading. Psycho-physical experiments show that infants (age 1–2 months) tend to fixate around an arbitrary single distinctive feature of the stimulus, like the corner of a triangle (Haith *et al.*, 1977; Salapatek and Kessen, 1973). Moreover, when presented with line drawings, children up to age 3–4 spend most of their time dwelling only on the internal details of a figure, and in general, children make more eye movements and are less likely than adults to look directly at a matching target in their first eye movements (Cohen, 1981). In comparison, adults display a strong tendency to look directly at forms that are informative, unusual, or of particular functional value (Antes, 1974; Loftus and Mackworth, 1978). Thus, it seems that gaze control in adults is indeed task and context dependent, but it is probably based on natal (hardwired) low-level local and context-free attentional mechanisms. At first, only the low-level context-free mechanisms are available. Gradually, as more information regarding the environment is being learned, higher-level processes take their place.

Active vision definitely needs high-level context-dependent attentional algorithms, but these should be adaptive trainable algorithms based on acquired knowledge, that use lower-level context-free attentional modules. Considering the fact that this research area is rather new, robust and efficient low-level attentional algorithms are the basic building blocks for machine visual attention.

2.2 MECHANISMS OF FOVEATION

Foveation in *active vision* is primarily motivated by biological systems, and thus, it is most natural to imitate the biological vision systems. The diversity of these systems is enormous (Vallerga, 1994), and in fact, every existing implementation of acquisition device has its biological counterpart.

The main approaches that are being used in computer vision for implementing foveation (in the sense of non-uniform resolution across the visual field) are the following:

- A single camera mounted on a moving device. The camera acquires a non-uniformly scanned image that is typically a log-polar one (Rojer and Schwartz, 1990; Biancardi *et al.*, 1993; Yamamoto *et al.*, 1995).
- Two monocular cameras configuration, consisting of a narrow-field-of-view high-resolution one, and a wide-field-of-view low-resolution one. In this configuration, the large field of view is used to received general information on the environment, and the narrow field of view is directed at regions of interest (Dickmanns and Graefe, 1988).
- Omni-directional lens with electronic foveation. In this configuration, a very large field of view is acquired using a fish-eye lens, and the image is projected on a regular CCD device. Once a region of interest is defined, however, the defined region could be sampled in high resolution by electronically correcting the image (Zimmermann and Kuban, 1992).

2.3 ATTENTIONAL MECHANISMS IN COMPUTER VISION

Attention in computer vision systems could be either *overt*, guiding the gaze, or *covert*, selecting a ROI within the captured image. In both cases, algorithms that analyse the image and select a region in it should be developed.

The first step in practically every artificial vision system consists of some form of edge detection. This step, which follows in general the early stages of biological systems, leads to the simplest mechanism for selecting ROI – looking for areas with large amount of edges. An early attentional operator based on grey-level variance (Moravec, 1977) is still being widely used in many systems. This operator is closely related to edge-based algorithms, since edges correspond to the variance in grey level. Other researchers suggested to measure *busyness* – the smoothed absolute value of the Laplacian of the data (Peleg *et al.*, 1987), rapid changes in the grey levels (Sorek and Zeevi, 1988), or smoothness of the grey levels (Milanese *et al.*, 1993). All of these methods try to measure, directly or indirectly, the density of edges, regardless of their spatial configuration.

If spatial relations of edges are to be considered, then, following early psychophysical findings (Attneave, 1954; Kaufman and Richards, 1969), interest points can be regarded also as points of high curvature of the edge map (Lamdan *et al.*, 1988; Yeshurun and Schwartz, 1989). Similarly, one could look for edge junctions (Brunnstrome *et al.*, 1992).

Regions of interest could also be defined by more specific features, or even by specific attributes of objects. Blobs, for example, are usually associated with objects that might attract our attention (Lindberg, 1993). Specific features are looked for in the application of fingerprints identification (Trenkle, 1994), detection of flaws in artificial objects (Magge *et al.*, 1992), and detection of man-made objects (characterized by straight lines and corners) in general images (Lu and Aggarwal, 1993). A method to direct attention to general objects is suggested in Casasent (1993).

Other visual dimensions that could be used as the core of attentional mechanisms are depth, texture, motion and color. Vision systems are either monocular, and then regions of interest could be defined as the regions where focused objects exist, or binocular, and then the horopter and the intersection of the optical axes could be used as the relevant cues (Olson and Lockwood, 1992; Theimer *et al.*, 1992; Coombs and Brown, 1993). Color as an attentional cue is suggested in Draper *et al.* (1993), and motion in Torr and Murray (1993).

Any attentional mechanism, and especially edge and feature based ones, should also consider the issue of the right scale for search. Since it is impossible to determine this scale in advance, a multi-scale representation should be preferred. This issue is discussed in Lindberg (1993), where existence of image features in multiple scales can enhance the probability of selecting a specific area for detailed analysis.

Most attentional mechanisms for computer vision applications are rather new, and thus their performance on natural images is not well studied and compared. In order to assess the performance of various attentional algorithms, one of the systems that have been recently developed (Baron *et al.*, 1994), includes an attentional module that could be easily interchanged, thus enabling a convenient testbed for benchmarks of such algorithms under the same environmental conditions.

Recently, symmetry has been suggested as a broader concept that generalizes most of the existing context-free methods for detection of interest points. Natural and artificial objects often give rise to the human sensation of symmetry. Our sense of symmetry is so strong that most man-made objects are symmetric, and the Gestalt school considered symmetry as a fundamental principle of perception. Looking around us, we get the immediate impression that practically every interesting area consists of a qualitative and generalized form of symmetry. In computer vision research, symmetry has been suggested as one of the fundamental non-accidental properties, which should guide higher-level processes. This sensation of symmetry is more general than the strict mathematical notion. For instance, a picture of a human face is considered highly symmetric by the layman, although there is no strict reflectional symmetry between both sides of the face.

Symmetry is being widely used in computer vision (Davis, 1977; Nevatia and Binford, 1977; Blum and Nagel, 1978; Brady and Asada, 1984; Atallah, 1985; Bigun, 1988; Marola, 1989; Xia, 1989; Schwarzinger *et al.*, 1992; Zabrodsky *et al.*, 1992). However, it is mainly used as a means of convenient shape representation, characterization, shape simplification, or approximation of objects, which have been already segmented from the background. A schematic (and simplified) vision task consists of edge detection, followed by segmentation, followed by recognition. A symmetry transform is usually applied after the segmentation stage. The symmetry transform suggested by (Reisfeld *et al.*, 1995) is inspired by the intuitive notion of symmetry, and assigns a *symmetry magnitude* and a *symmetry orientation* to every pixel in an image at a low-level vision stage which follows edge detection. Specifically, a *symmetry map* is computed, which is a new kind of an edge map, where the magnitude and orientation of an edge depends on the symmetry associated with the pixel. Based on the magnitude and orientation of the symmetry map, it is possible to compute *isotropic* symmetry, and *radial* symmetry, which emphasizes closed contours. Strong symmetry edges are natural interest points, while linked lines are symmetry axes. Since the symmetry transform can be applied immediately after the stage of edge detection, it can be used to direct higher-level processes, such as segmentation and recognition, and can serve as a guide for locating objects.

In Figure 2.1 we demonstrate the problem that might arise by using a simple edge-density-based attentional mechanism. In this image, the density of the edge map is high in too many locations, and thus it would be rather a difficult task

Figure 2.1   Natural image with high edge density. Top: original image and the peaks of the radial symmetry. Bottom (left to right): edge detection, isotropic symmetry and radial symmetry.

to detect areas of interest by a simple edge map. However, if the spatial configuration of the edges is considered, as is done by using the generalized symmetry transform, more meaningful regions will be attended to. Operation of the symmetry transform on another natural image (Figure 2.3) is also demonstrated. In this case, other attentional operators might also be useful, since the targets could be detected by considering the blobs in the image. The scale issue could be demonstrated in Figure 2.2. In this image, which is obtained by zooming in on Figure 2.1, we demonstrate that the same algorithm with the same parameters could be used in different scales.

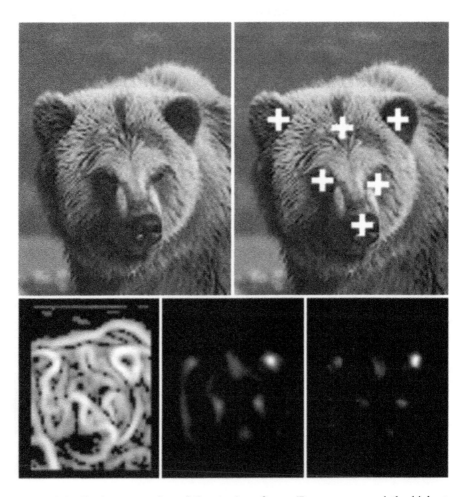

Figure 2.2 Further processing of the previous figure. Top: area around the highest radial symmetry peak in finer resolution and the peaks of the radial symmetry marked by crosses. Bottom (left to right): edge detection, isotropic symmetry and radial symmetry.

Figure 2.3 Detection of objects in a natural scene. The crosses mark the highest peaks of the radial symmetry map.

2.4 CONCLUSIONS

Machine attention should emerge from low-level mechanisms that detect visual cues at early stages, as well as from mechanisms that detect specific objects. There is however much more to attention than simple context-free mechanisms. As we have argued, attention, like almost every other visual task, is trapped in the local-global, top-down bottom-up, and context-free context-dependent vicious circle. Attention probably requires some assumptions about the environment, since it is the deviation from the expected framework and irregularities that should draw the attention of the system. This point is obviously not easily approached in computer vision, but some attempts to formalize it have been made (Zucker *et al.*, 1975; Gong, 1992).

A possible path to follow should consist of a three-level paradigm:

- A context free direct computation of a set of simple and early mechanisms, like color, motion, edge density or the generalized symmetry.
- An analysis of the geometry of this early map, based on general, and task dependent knowledge.
- A conventional object detection and recognition performed only in locations indexed by the previous stages.

This approach is not necessarily bottom up, as it seems to be on first sight, since the specific set of context-free features used in the lower level could be selected and modified by the higher levels, once a specific task is performed.

The approach could be demonstrated, for example, in the task of detecting persons and facial features in images by using the *generalized symmetry transform*. First, the context-free generalized symmetry map is computed, and then one might look for a geometric pattern where the symmetry peaks are vertically arranged, below a circular symmetry peak (persons), or three symmetry peaks that form a triangle (facial features). Indexed locations could then be specifically analysed by edge based or intensity based recognition schemes.

Attention is among the most complex psychological phenomena that are being studied (Cantoni *et al.*, 1996), and thus, it is only the first and most superficial aspects of it that are being considered in computer vision. But as we become more and more aware of the usefulness of considering biological approaches in artificial systems, the role of attention there could not be overestimated.

REFERENCES

Aboot, A. and Ahuja, N. (1988) Surface reconstruction by dynamic integration of focus, camera vergence and stereo. *Proc. of the 2nd Int. Conf. on Computer Vision.*

Aloimonos, J., Weiss, I. and Bandyopadhyay, A. (1987) Active vision. *Int. Journal of Computer Vision*, 334–356.

Antes, J. (1974) The time course of picture viewing. *Journal of Experimental Psychology* **103**, 62–70.

Atallah, M. (1985) On symmetry detection. *IEEE Transactions on Computers* **C-34**, 663–666.

Attneave, F. (1954) Informational aspects of visual perception. *Psychological Review* **61**, 183–193.

Bajcsy, R. (1988) Active perception. *Proceedings of the IEEE* **76**(8), 996–1006.

Ballard, D. (1990) Animated vision. *Technical Report TR 61*, University of Rochester, Department of Computer Science.

Baron, T., Levine, M. and Yeshurun, Y. (1994) Exploring with a foveated robot eye system. *Proc. 12th IAPR Int. Conf. on Pattern Recognition*, 377–380.

Biancardi, A., Cantoni, V. and Lombardi, L. (1993) Computer vision systems: functionality and structure integration. In: *Intelligent perceptual systems* (ed. V. Roberto), 70–83. Springer-Verlag, Berlin, Germany.

Bigun, J. (1988) Pattern recognition by detection of local symmetries. In: *Pattern recognition and artificial intelligence* (eds E. Gelsema and L. Kanal), 75–90. Elsevier North Holland, Amsterdam, The Netherlands.

Blum, H. and Nagel, R. (1978) Shape description using weighted symmetric axis features. *Pattern Recognition* **10**, 167–180.

Brady, M. and Asada, H. (1984) Smoothed local symmetries and their implementation. *The Int. Journal of Robotics Research* **3**(3), 36–61.

Brunnstrome, K., Lindeberg, T. and Eklundh, J. (1992) Active detection and classification of junctions by foveation with a head-eye system guided by the scale-space primal sketch. *Proc. of the 2nd Eur. Conf. on Computer Vision*, 701–709, S. Margherita, Italy.

Cantoni, V., Caputo, G. and Lombardi, L. (1996). Attentional engagement in vision systems, *this volume*.

Casasent, D. (1993) Sequential and fused optical filters for clutter reduction and detection. *Proceedings of the SPIE* 1959, 2–11.

Clark, J. and Ferrier, N. (1988) Modal control of an attentive vision system. *Proc. of the 2nd Int. Conf. on Computer Vision*, 514–519.

Cohen, K. (1981) The development of strategies of visual search, in eye movements. In: *Cognition and visual perception* (eds D. Fisher, R. Monty and J. Senders), 299–314. Erlbaum, Hillsdale, NJ.

Coombs, D. and Brown, C. (1993) Real-time binocular smooth pursuit. *Int. Journal of Computer Vision* 11, 147–164.

Crowley, J. (1991) Towards continuously operating integrated vision systems for robotics applications. *Proc. SCIA-91, 7th Scandinavian Conf. on Image Analysis*, Aalborg, Denmark.

Culhane, S. and Tsotsos, J. (1992) An attentional prototype for early vision. *Proc. of the 2nd Eur. Conf. on Computer Vision*, 551–560, S. Margherita, Italy.

Davis, L. (1977) Understanding shape: I. Symmetry. *IEEE Trans. on Systems, Man, and Cybernetics*, 204–211.

Dickmanns, E. and Graefe, V. (1988) Applications of dynamic monocular machine vision. *Machine Vision and Applications* 1, 241–261.

Draper, B. A., Buluswar, S., Hanson, A. R. and Riseman, E. M. (1993) Information acquisition and fusion in the mobile perception laboratory. *Proceedings of the SPIE* 2059, 175–187.

Gong, S. (1992) Visual behaviour: modelling 'hidden' purposes in motion. *Proceedings of the SPIE* 1766, 583–593.

Haith, M., Bergman, T. and Moore, M. (1977) Eye contact and face scanning in early infancy. *Science* 198, 853–855.

Kaufman, L. and Richards, W. (1969) Spontaneous fixation tendencies for visual forms. *Perception and Psychophysics* 5(2), 85–88.

Lamdan, Y., Schwartz, J. and Wolfson, H. (1988) On recognition of 3-d objects from 2-d images. *Proc. of IEEE Int. Conf. on Robotics and Automation*, 1407–1413.

Lindberg, T. (1993) Detecting salient blob-like image structures and their scales with a scale-space primal sketch: a method for focus-of-attention. *Int. Journal of Computer Vision* 11, 283–318.

Loftus, G. and Mackworth, N. (1978) Cognitive determinants of fixation location during picture viewing. *Human Perception and Performance* 4, 565–572.

Lu, H. and Aggarwal, J. (1993) Applying perceptual organization to the detection of man-made objects in non-urban scenes. *Pattern Recognition* 25, 835–853.

Magge, M., Seida, S. and Franke, E. (1992) Identification of flaws in metallic surfaces using specular and diffuse bispectral light sources. *Proceedings of the SPIE* 1825, 455–468.

Marola, G. (1989) On the detection of the axis of symmetry of symmetric and almost symmetric planar images. *IEEE Trans. Pattern Analysis and Machine Intelligence* 11(1), 104–108.

Milanese, R., Pun, T. and Wechsler, H. (1993) A non-linear integration process for the selection of visual information. In: *Intelligent perceptual systems* (ed. V. Roberto), 322–336. Springer-Verlag, Berlin, Germany.

Moravec, H. (1977) Towards automatic visual obstacle avoidance. *Proc. of IJCAI*, 584–590.

Nevatia, R. and Binford, T. (1977) Description and recognition of curved objects. *Artificial Intelligence* 8, 77–98.

Olson, T.J. and Lockwood, R.J. (1992) Fixation-based filtering. *Proceedings of the SPIE* 1825, 685–695.

Peleg, S., Federbush, O. and Hummel, R. (1987) Custom made pyramids. In: *Parallel computer vision* (ed. L. Uhr), 125–146. Academic Press, New York, NY.

Reisfeld, D., Wolfson, H. and Yeshurun, Y. (1995) Context free attentional operators: the generalized symmetry transform. *Int. Journal of Computer Vision* **14,** 119–130, special issue on qualitative vision.

Rimey, R. and Brown, C. (1992) Where to look next using a Bayes net: incorporating geometric relations. *Proc. of the 2nd Eur. Conf. on Computer Vision*, 542–550, S. Margherita, Italy.

Rojer, A. and Schwartz, E. (1990) Design considerations for a space-variant visual sensor with complex logarithmic geometry. *Proc. of the 10th IAPR Int. Conf. on Pattern Recognition*, 278–285.

Salapatek, P. and Kessen, W. (1973) Prolonged investigation of a plane geometric triangle by the human newborn. *Journal of Experimental Child Psychology* **15,** 22–29.

Schwarzinger, M., Zielke, T., Noll, D., Brauckmann, M. and von Seelen, W. (1992) Vision-based car following: detection, tracking, and identification. *Proc. of the IEEE Intelligent Vehicles '92 Symp.*, 24–29.

Sorek, N. and Zeevi, Y. (1988) Online visual data compression along a one-dimensional scan. *Proceedings of the SPIE*, 1001, 764–770.

Theimer, W.M., Mallot, H.A. and Tolg, S. (1992) Phase method for binocular vergence control and depth reconstruction. *Proceedings of the SPIE* 1826, 76–87.

Tistarelli, M. and Sandini, G. (1990) Estimation of depth from motion using an anthropomorphic visual sensor. *Image and Vision Computing* **8**(4), 271–278.

Torr, P. and Murray, D. (1993) Statistical detection of independent movement from a moving camera. *Image and Vision Computing* **11,** 180–187.

Trenkle, J.M. (1994) Region of interest detection for fingerprint classification. *Proceedings of the SPIE* 2103, 48–59.

Vallerga, S. (1994) The phylogenetic evolution of the visual system. In: *Human and machine vision: analogies and divergencies* (ed. V. Cantoni), 1–13. Plenum Press, New York, NY.

Xia, Y. (1989) Skeletonization via the realization of the fire front's propagation and extinction in digital binary shapes. *IEEE Trans. Pattern Analysis and Machine Intelligence* **11**(10), 1076–1089.

Yamamoto, H., Yeshurun, Y. and Levine, M. (1995) An active foveated vision system: attentional mechanisms and scan path convergence measures. *Journal of Computer Vision, Graphics and Image Processing*, in press.

Yeshurun, Y. and Schwartz, E. (1989). Shape description with a space-variant sensor: algorithm for scan-path, fusion, and convergence over multiple scans. *IEEE Trans. Pattern Analysis and Machine Intelligence* **11**(11), 1217–1222.

Zabrodsky, H., Peleg, S. and Avnir, D. (1992). A measure of symmetry based on shape similarity. *Proc. CVPR*, Urbana, Illinois.

Zimmermann, S. and Kuban, D. (1992) A video pan/tilt/magnify/rotate system with no moving parts. *Proc. IEEE-AIAA 11th Digital Avionics Systems*, 523–531.

Zucker, S., Rosenfeld, A. and Davis, L. (1975) General purpose models: expectations about the unexpected. *Proc. of the 4th IJCAI*, 716–721.

Zucker, S., Dobbins, A. and Iverson, L. (1989) Two stages of curve detection suggest two styles of visual computation. *Neural Computation* **1,** 68–81.

3

The Active Vision Idea

3.1 WHAT IS 'ACTIVE'?

The word 'active' naturally appears as opposite to 'passive', and it carries with itself the idea of 'action'. The point is: who is the actor? And what is acted on? The purpose of a vision system, be it natural or artificial, seems to be the perception of an external physical scene without implying any influence on it. In other words, we could think that letting our eyes open, or acquiring data from a camera, and processing the data so collected, is exactly what we need to visually 'perceive' the world. We will discuss later on these sufficient conditions. Now we want to clarify from the beginning that for *active* vision we indicate the approach followed by a system that manages its own resources in order to simplify a perceptual task. The resources on which the observer builds its perceptual abilities comprise visual sensors and processing devices, but also other types of sensors, actuators and even illumination devices, if needed. It is important to distinguish between active sensing and active vision. A *sensor* is considered active if it conveys some sort of energy into the external environment in order to measure, through the reflected energy, some physical property of the scene. Examples of these are laser scanners or ultrasonic range finders. An active *observer* may use either passive or active sensors, or both, but the concept of activity we are introducing relates to the observer, not to the sensor. The active observer influences acquisition parameters like camera position and orientation,

ARTIFICIAL VISION   Copyright © 1997 Academic Press Ltd
ISBN 0-12-444816-X   All rights of reproduction in any form reserved

lens focus, aperture and focal length, and sampling resolution. The active observer manages also its computational resources by devoting computing power and processes to critical sub-tasks or processes useful for reaching a goal. In other words, active vision has to deal with *control* of acquisition parameters, with selective *attention* to phenomena that take place in the spatio-temporal surroundings, and with *planning* of activities to perform in order to reach a goal. The main point here is that an intelligent observer should make use of all its resources in an economical and integrated way to perform a given task. By taking this point of view, as we do, we cannot consider vision any more as an isolated function.

3.2 HISTORICAL FRAME

The history of research in computer vision starts, together with the history of artificial intelligence and, ultimately, with the increasing availability of computing resources and languages, at the beginning of the 1960s. The active vision paradigm however has been made explicit and has been spreading in the scientific community only since the second half of the eighties. Why? There are several reasons, but we can group them into two classes: technological limitations and cultural influences.

Let us start with technological limitations. Useful computer vision has to face the difficulties posed by non-trivial, real-world images. One image provides a significant amount of data, that becomes enormous if we have to deal with an image sequence in a situation that involves motion, and we want to approach real-time performance. Moreover, anything can happen in an image; not only noise, but the extreme variability of the real world constitutes a very difficult problem at the computational theory and algorithmic levels (Marr, 1982). So, in search for a solution, a heavier and heavier burden is laid on the available computing resources. For these reasons the past predominant approach considered a static observer trying to recognize objects by analysing monocular images or, at most, stereo pairs. Motion was seen as a difficult problem that was wise avoiding whenever possible, like, for example, in object recognition and scene understanding tasks in a static environment. It was thought that observer motion would have further complicated these already difficult tasks. This intuition would have been proved wrong.

Beyond that, acting on camera extrinsic and intrinsic parameters posed technical and cost problems connected with the limited availability of light-weight TV cameras, efficient and easy to control motor systems, motorized lenses, fast acquisition devices and communication channels, and multi-processor systems.

The other class of reasons for not considering motion stemmed from the origin of computer vision, rooted in symbolic artificial intelligence. Inspired by the traditional AI perspective, vision was then mostly seen as an input functionality for the higher-reasoning processes. In this way the visual act resulted as decomposed in successive layers of computations, organized in an open loop hierarchy.

At the lowest level were processes devoted to noise reduction, image enhancement, edge-points extraction, stereo disparity calculation, etc. At an intermediate level were processes that extracted and organized feature sets from one or more intrinsic images. At the higher level these feature sets were compared with internal models in order to recognize objects and to reason about the physical reality that produced the image(s) and that had been 'reconstructed' by the previous processes.

Unfortunately, the task revealed itself as being much more difficult than expected. Even if, at the beginning of research in artificial vision, the undeclared acceptance of the naive idea that vision was so easy and effortless for us that it would have been almost the same for computers was conceivable, it shortly disappeared under the disillusions of the first attempts in dealing with real image analysis tasks.

One of the reasons that make vision unpenetrable is the fact that the important processes that give rise to perception in humans escape the domain of introspection. So, while symbolic AI has been inspired by the logical paradigm of thought, it has been very soon clear that the same might not satisfactorily hold for perception.

One could ask if it has been satisfactory or successful at all for AI itself; but vision goes beyond the difficulties shared with AI by adding the puzzling involvement of the so-called 'primary processes' that account for perceptual organization and interpolation. As gestaltist researchers have pointed out (Kanizsa, 1979), the primary processes themselves have an active character.

Among the merits of the active approach to computer vision is the recognition of the fact that visual capabilities should be considered and designed as strictly integrated within the global architecture of an intelligent autonomous system.

The very same control loop of perception and action, typical of the active vision approach, seems to point toward a more strict relationship between 'vision' and 'thought'. The same relationship had been previously recognized by psychologists that refused the artificially sharp separation between the two concepts (Kanizsa, 1979). The active approach upholds vision to confront artificial intelligence on a parithetic level, by recognizing the importance of processes like categorizing, hypothesis formation and testing, and action planning. This approach to computer vision has emerged after the difficulties encountered by the reconstructionist school, flourished in the eighties.

3.3 THE RECONSTRUCTIONIST SCHOOL

The reconstructionist school, which has seen its highest expression in the influential work by Marr (1982), has neglected for a long time the importance of the observer as being immersed in the scene and interacting with it in ways much more complex than the simple acquisition of images. Nevertheless, an important part of Marr's contribution, perhaps the most important one, is the rigorous computational approach to vision. In this perspective he stated a clear subdivision among three levels in the solution of a problem: the computational theory, the algorithmic level and the implementation.

In the work by Marr and the other researchers that followed his approach, attention is explicitly focused on *what* is present in a scene and *where* it is, and on the *process* for extracting this information from images. To know what is present in a scene means to recognize objects, and the most distinctive characteristic that allows to perceptually discriminate between different objects is shape.

To define what is shape is difficult, but it is easy to accept its relationship with the topological and geometrical properties of the object. In the same way, to know where an object is can mean to know its geometrical relationship with the three-dimensional world in which it is placed. Thus, as obvious consequence, we see the importance given to the problem of obtaining a good geometrical representation of the viewed scene. This approach has been called *reconstructionist* because it tries to reconstruct the three-dimensional structure of the scene from its two-dimensional projections on images.

It has given rise to many sophisticated algorithms sometimes collectively indicated as 'shape from X', where 'X' can be for example instantiated by 'shading', 'contour' or 'texture'. These techniques recover the local three-dimensional shape of the surface of an object from two-dimensional image cues.

Problems arise from several factors, some of which are:

- The image formation process itself, that is the projection on a two-dimensional surface of a three-dimensional scene thus entailing loss of depth information.
- The uncertainty about the number and direction of the illuminants.
- The variability of reflectance characteristics of objects surfaces.
- The complex interaction among local reflectance and surface orientation and illumination and observation directions.

Due to these and other aspects many shape from X problems suffer from ill-posedness, nonlinearity, instability or a combination of them.

3.4 ACTIVE RECONSTRUCTION

In a series of seminal papers – the first presented in 1987 – Aloimonos and others (Aloimonos *et al.*, 1987; Aloimonos *et al.*, 1988; Aloimonos and Shulman, 1989) demonstrated how a controlled motion of the sensor can considerably reduce the complexity of the cited classical vision problems that fall into the 'Shape-from-X' group.

For all the three cases cited above, the authors showed that adequately controlled observer motion results in a mathematical simplification of the computations.

Let us take for example shape from shading, that is, a process to infer local orientation of object surfaces from apparent luminance values (Horn, 1977; Horn, 1986). It results that if we want to solve it with a single monocular view, we need to take additional assumptions about the smoothness and uniformity in reflectance characteristics of the surface, and the uniformity of lighting. These often unrealistic assumptions limit the applicability of techniques and algorithms based on them, but they must be adopted to regularize a generally underconstrained, ill-posed problem like passive shape from shading. Things become easier with two images in stereo configuration, leaving aside the correspondence problem, and even better with three images, but the problem is still nonlinear. Moreover stereo images pose two different kinds of problems in short and long baseline configurations: a short baseline makes the correspondence problem easier and allows for a linearization of the equations to solve by taking only up to the linear term of a Taylor series expansion, but the resulting accuracy is low; a long baseline reaches good accuracy but suffers from a difficult correspondence problem that can trap the search for solution in a local minimum.

Aloimonos and colleagues presented an active technique that takes the advantages of both stereo configurations. This technique decouples the process of finding the unknowns (surface orientation and reflectance parameters, plus lighting distribution) at every single point in the image. The adopted reference system is fixed. The camera moves by shifting the position of the optical centre in the XY plane, that is parallel to the image plane. For the mathematical details the reader can refer to one of the papers (Aloimonos *et al.*, 1987; Aloimonos *et al.*, 1988; Aloimonos and Shulman, 1989).

The effect of camera motion is that more data are collected in a controlled way such that the equations for the different surface points become coupled along time (subsequent camera positions) and are no more coupled for adjacent points on the surface.

This gets rid of the old unrealistic assumptions on maximal surface smoothness and uniform reflectance and lighting. In this way it is possible to separate shape information from reflectance and lighting variations, thus detecting and measuring all of them. Mathematical and computational advantages of an active approach to other shape-from-X modules have been similarly shown.

Given the under constrained nature of problems of shape reconstruction from intensity images, the strength of the active approach lies in substituting unrealistic constraints, artificially imposed on the nature of the outer scene, with actual and perfectly known constraints given by controlled observer motion.

The main task of model based object recognition, primarily targeted by shape from X techniques, has been often accompanied by automatic model extraction, in which the same techniques for surface reconstruction were used for building a CAD-like model of an unknown object. Chapter 4 of this book by E. Trucco gives a comprehensive treatment of both these tasks, along with an extensive bibliography.

We add here that the possibility of a fine control of camera parameters can conquer also problems given by difficult shapes like tori (Kutulakos and Dyer, 1994), and that the planning of multiple observations gives great advantages for efficiency of object recognition techniques (Gremban and Ikeuchi, 1994; Hutchinson and Kak, 1989). These advantages can be summarized into the acquired abilities of eliminating ambiguities and of incrementally gathering several cues that contribute to the solution of the problem at hand.

3.5 PURPOSIVE VISION

Object *recognition* constitutes only one of the major visual problems of interest, but other activities build on information obtained through vision. The most important one is *navigation*, that is, moving around in the world with the help of visual sensors. Moreover there are many problems or subproblems, that an intelligent, autonomous, perceiving agent must frequently solve, that need only some specific, task dependent information to be extracted from images. Some of these problems are: obstacle detection and avoidance, detection of independent motion, tracking of a moving object, interception, hand-eye coordination, etc. Trying to solve each of these problems separately can lead to much more economical solutions and to the realization of effective working modules to be activated on demand when need arises (Aloimonos, 1990; Ikeuchi and Hebert, 1990). The alternative approach now sketched has been called *purposive vision*, referring to the utilitarian exploitation of both system resources and surrounding world information, or *animate vision*, with approximately the same meaning (Ballard and Brown, 1992).

We have already seen that an active approach can help in the recovery of the three-dimensional structure of the observed scene. Actually, even if the geometrical structure could be extracted through an analysis of image sequences, under controlled observer motion conditions, the practical implementation of a working structure from motion module remains difficult. This module would grant us with solutions to both the object recognition and the navigation problems, but the use we make of our visual system may not always entail complete scene reconstruction. Many tasks that make use of visual input need only to extract some very specific information. This means that not every visual information nor every part or object present in a scene should be respectively extracted or analysed in detail, but only the single elements that are currently useful for the completion of the tasks at hand.

This approach has led some researchers to develop specific algorithms for solving specific sub-tasks like for example: detection of independent motion by a moving observer, tracking, estimation of relative depth (Ballard and Brown, 1992; Huang and Aloimonos, 1991; Fermuller and Aloimonos, 1992; Sharma and Aloimonos (1991)). Tracking, in particular, has attracted the attention of many researchers because of its importance (Fermuller and Aloimonos, 1992; Coombs and Brown, 1991; the first seven papers in Blake and Yuille, 1992). There are different reasons for such attention. By 3D-tracking of a moving object, the observer can keep it in sharp focus at the center of the image, in order to collect as much information as possible about it.

In the same time it is easier to isolate the tracked object from the background through motion blur of the second one. The tracking motion parameters of the observer give direct information on relative position and speed, for example for interception tasks. For the same reason, tracking of a fixed environment point in the observer image space gives information on observer trajectory and speed, extremely useful for navigation. There is another important aspect to consider: tracking or fixation establishes a relationship between the observer and the fixated object reference frames (Ballard and Brown, 1992). This helps in translating observer centered descriptions extracted from the images into object centered descriptions stored as models, thus making the path to model based object recognition easier.

Now, some questions arise. If there is convenience in selecting the information to be extracted from the scene, what kind of information is best suited to the task and to the resources at hand? How to combine sensing with action, vision with behaviour? How much explicit representation do we need? And which kind of representations?

3.6 INTELLIGENT PERCEPTION AND THE ACTIVE APPROACH

By taking a systemic view of perception and perceiving agents, we should not overlook any of the 'subsystems' that contribute to the perceptual task. Let us consider a visual system that surely works: our own. We constantly move our eyes, head and body to act in the world, and we have other sensors, in our ears and muscles, that provide information about our *movements*; so the observer motion must not be such an obstacle on the road to vision; it is probably more an advantage. For sure it is an effective way to collect an amazing amount of useful information, while dominating the spatio-temporal huge dimension of the input data, by sequentially directing the fovea only on small regions of interest in the scene. This paradigm allows to separate the *what* and *where* parts (Marr, 1982) of the classical image understanding problem.

Another aspect of the human visual system that has attracted the researchers is the foveated eye. By distributing the single sensitive elements with varying resolution it is possible to keep the data rate at levels that are orders of magnitude lower than those of uniformly sampling sensors (Schwartz, 1977; Sandini and Tagliasco, 1980), while keeping the same high level of detail in the fixated region of interest. Again, these new sensors would be of little use without active gaze control capabilities.

There is an aspect of image understanding that deserves careful consideration. As shown by (Yarbus, 1967) the visual exploration of a picture by a human observer follows different paths, according to the particular task he/she has been assigned with. Not only different global tasks may cause the exploration of different regions in the image, but some regions that are of particular significance for the given task are analysed more carefully; gaze is positioned on them with insistence, often coming back to these points of interest after a round trip to other regions. These experiments suggest two conclusions:

- What is important is to collect, in as short a time as possible, the information that allows to perform the task assigned to the observer, leaving aside the rest.
- To answer some questions about the scene one needs to analyse few restricted areas for a long time, meaning perhaps that the extraction of certain kinds of information is difficult compared to the extraction of others.

This view is not completely new, and the idea that a complete internal representation of the world might not be the best approach to the vision problem can be traced for example in the ecological approach proposed by Gibson (1950, 1966, 1979). According to him, the world around us acts as a huge external repository of the information necessary to act, and we directly extract from time to time the elements that we need.

3.7 ACTIVE COMPUTATIONAL VISION AND REPRESENTATIONS

The considerations just presented in the previous paragraph introduce us into the other problem that we wanted to address: *representations*. Do we need them? And, if yes, at what extent? And of which kind? The ecological approach seems to suggest a negative answer to the first of these questions; and the work by Brooks (1987) in the track of behaviorism follows this approach. Conversely, representations play a very important role in the Marr paradigm. This is mainly due to the rigorous computational approach followed by him. In this computational perspective the role of representations is correctly emphasized as that of formal schemes that make explicit some information about an entity. Moreover, formal schemes are the only objects that can be manipulated and transformed, one into the other, by computer programs. It should also be noticed that even the extreme position of Brooks against representation is substantially against the use of certain kinds of it, and even his robots make use of representations. The point is that there exist many forms of representation that are influenced by the available hardware, both for information gathering (sensors) and processing (neurons, transistors). Going back to the computational vision approach, the visual process, or at least its prominent part, is seen as transforming an image into a different kind of representation. As image, or input representation, is generally taken a regularly spaced grid of pixels, each of which carrying a numerical measure of the irradiance, taken at the corresponding point on the sensor. In this way what is made explicit by the representation is the measure of light, while other information, such as reflectance of the surface, texture, shape, relative position and motion of the object are left in the background. The effort is then to devise a process, or a collection of processes, that extracts interesting information by transforming the input representation into a new one that make that information explicit. Given the enormous difficulties often encountered on this path, we could ask ourselves if the addition of some different kind of input signal could help.

And how should the output representation be organized? What kind of information, embedded in the input, should be made explicit? The answer depends on the task for which visual processing is performed. The suitability of a representation for a given task depends on whether it makes explicit and easy to use the information relevant to the task itself. The traditional reconstructionist approach gave absolute preference to shape and spatial relations. If shape is the key point, then the final output representation has been selected as a three-dimensional model of the observed scene. Model based recognition uses internal geometrical representations, often of the same kind of those used in CAD systems, to be matched against the resulting representation produced by the visual processing.

It is clear that a certain amount of representation is needed, at least because today we are forced to use digital computers and software programs. But it is not necessarily the case that these representations must resemble the geometrical appearance of the observed scene. We agree that the key point is space representation, but it seems reasonable to suggest the role of eye movements and gaze control into an internal representation of space. Just think about what happens when somebody throws a ball to you. The key task is visually tracking the ball, foreseeing its trajectory in 3D space and organizing and activating muscle commands in order to intercept it. During that time interval CAD-like 3D scene reconstruction has no meaning; the spatial relationship between you and the ball is really the only important parameter. But the tight integration in the same control loop of ball detection and localization in the image and oculo motor commands, that allow for fixation and tracking of the ball, represents the most economical solution of the task (Coombs and Brown, 1991; Brown, 1990). Vergence commands give the distance of the ball at every instant, and they can also represent it. Gaze control commands give the direction in space from the observer to the ball, and, similarly, they can represent it, along with velocity information that is necessary for interception.

If internal motion commands can play a significant role in 'representing' the dynamic situation of tracking and interception (Berthoz, 1993) there is no apparent reason why they could not take an active role in space representation in general. The spontaneous exploration carried out by any normal individual introduced for the first time into an unknown environment – a room, for example – can be reasonably interpreted as the process of building a model of the space in which he/she is situated. But a satisfying answer to the question if this representation entails 3D-surface models, or internal exploration movements commands, or both, is far from being here.

3.8 CONCLUSIONS

Active vision is one of the hot subjects of research in the field of computer vision (Aloimonos, 1992; Swain, 1994; Fiala *et al.*, 1994). This approach has already proved its usefulness in solving some hard problems in computer vision, but it has demonstrated its power also in complex real-time applications (Dickmanns and Christians, 1989; Dickmanns *et al.*, 1990) and it is still in full development, so that more results are expected. It will not solve all the problems that still confront the researchers in this stimulating area, but it has already reached the result of revitalizing interest in the research community. Moreover, two important aspects of visual perception have been put in a different perspective: the position of vision, and perception in general, in the context of an intelligent autonomous agent, and the nature, role and meaning of representations. Vision should not be considered as an isolated function; rather it should be viewed as a powerful tool, so strictly integrated with the capabilities of an intelligent system that any artificial separation between vision and intelligence should be banned. These interactions will probably bring new insights and advantages in the study of both disciplines. In the same way active vision has the merit of having made explicit the relationship between perception and action, and the consequent importance of control (Clark and Ferrier, 1988; Rimey and Brown, 1994).

REFERENCES

Aloimonos, J. (1990) Purposive and qualitative active vision. *Proc. 10th Int. Conf. on Pattern Recognition*, Atlantic City, New Jersey.

Aloimonos, J. (ed.) (1992) *CVGIP: image understanding, special issue on purposive, qualitative, active vision*, **56**, 1, Academic Press.

Aloimonos, J. and Shulman, D. (1989) *Integration of visual modules: an extension of the Marr paradigm*. Academic Press, Boston, Massachusetts.

Aloimonos, J., Weiss, I. and Bandyopadhyay, A. (1987) Active vision. *Proc. 1st IEEE Int. Conf. on Computer Vision*, 35–54, London, UK.

Aloimonos, J., Weiss, I. and Bandyopadhyay, A. (1988) Active vision. *Int. Journal of Computer Vision* **1**, 333–356.

Ballard, D.H. and Brown, C.M. (1992) Principles of animate vision. *CVGIP Image Understanding* **56**, 1, 3–21.

Berthoz, A. (ed.) (1993) *Multisensory control of movement*. Oxford University Press, New York.

Blake, A. and Yuille, A. (eds) (1992) *Active vision*. MIT Press, Cambridge, Massachusetts.

Brooks, R. (1987) Intelligence without representation. *Proc. Workshop on the Foundations of AI*.

Brown, C.M. (1990) Gaze controls with interactions and delays. *IEEE Trans. on Systems, Man and Cybernetics* **20**, 3.

Clark, J.J. and Ferrier, N.J. (1988) Modal control of an attentive vision system. *Proc. 2nd Int. Conf. on Computer Vision*, 514–523, IEEE Press, Tampa, Florida.

Coombs, D.J. and Brown, C.M. (1991) Cooperative gaze holding in binocular robot vision. *IEEE Control Systems*, 24–33.

Dickmanns, E.D. and Christians, T. (1989) Relative 3D-state estimation for autonomous visual guidance of road vehicles. *Intelligent Autonomous Systems* **2**, 683–693.

Dickmanns, E.D., Mysliwetz, B. and Christians, T. (1990) An integrated spatio-temporal approach to automated visual guidance of autonomous vehicles. *IEEE Trans. on Systems, Man and Cybernetics* **20**, 1273–1284.

Fermuller, C. and Aloimonos, J. (1992) Tracking facilitates 3-D motion estimation. CAR-TR-618, Univ. of Maryland.

Fiala, J.C., Lumia, R., Roberts, K.J. and Wavering, A.J. (1994) TRICLOPS: A tool for studying active vision. *Int. Journal of Computer Vision* **12**, 2/3, 231–250.

Gibson, J.J. (1950) *The perception of the visual world*. Houghton Mifflin, New York.

Gibson, J.J. (1966) *The senses considered as perceptual systems*. Houghton Mifflin, New York.

Gibson, J.J. (1979) *The ecological approach to visual perception*. Houghton Mifflin, New York.

Gremban, K.D. and Ikeuchi, K. (1994) Planning multiple observations for object recognition. *Int. Journal of Computer Vision* **12**, 2/3, 137–172.

Horn, B.K.P. (1977) Understanding image intensities. *Artificial Intelligence* **8**, 201–231.

Horn, B.K.P. (1986) *Robot vision*. McGraw Hill, New York.

Huang, L. and Aloimonos, J. (1991) Relative depth from motion using normal flow: an active and purposive solution. CAR-TR-535, Univ. of Maryland.

Hutchinson, S. and Kak, A. (1994) Planning sensing strategies in a robot work cell with multi-sensor capabilities. *IEEE Trans. on Robotics and Automation* **5**, 6.

Ikeuchi, K. and Hebert, M. (1990) Task-oriented vision. *Proc. DARPA Image Understanding Workshop*, 497–507.

Kanizsa, G. (1979) *Organization in vision*. Praeger, New York.

Kutulakos, K.N. and Dyer, C.R. (1994) Recovering shape by purposive viewpoint adjustment. *Int. Journal of Computer Vision*, **12**, 2/3, 113–136.

Marr, D. (1982) *Vision*. Freeman, San Francisco, California.

Rimey, R.D. and Brown, C.M. (1994) Control of selective perception using Bayes nets and decision theory. *Int. Journal of Computer Vision* **12**, 2/3, 173–207.

Sandini, G. and Tagliasco, V. (1980) An anthropomorphic retina-like structure for scene analysis. *Comp. Vision, Graphics and Image Processing* **14,** 365–372.

Schwartz, E.L. (1977) Spatial mapping in the primate sensory projection: analytical structure and relevance to perception. *Biological Cybernetics* **25,** 181–194.

Sharma, R. and Aloimonos, J. (1991) Robust detection of independent motion: an active and purposive solution. CAR-TR-534, Univ. of Maryland.

Swain, M. (ed.) (1994) Special issue on active vision II. *Int. Journal of Computer Vision* **12,** 2/3.

Yarbus, A.L. (1967) *Eye movements and vision.* Plenum Press, New York.

4

Active Model Acquisition and Sensor Planning

4.1 INTRODUCTION

Active vision systems (Blake and Yuille, 1993) can reconfigure themselves in order to achieve the best conditions to carry out their perceptual task. Parameters to reconfigure include sensor position, focus of attention, intrinsic parameters of sensors and sensor motion. This chapter focuses on two tasks made possible by the controlled motion of an active observer: (a) automatic 3D model acquisition and (b) optimal, model-based active inspection. The former allows the observer to build a 3D geometric model incrementally, as an object is observed from different viewpoints; the latter uses the model to find optimal inspection positions for reconfigurable sensors. Notice that none of the two tasks could be performed by a passive observer, which could not alter the relative position of object and scene, and could not direct the sensors to any desired positions.

We shall assume that the active observer is equipped with *range* and *intensity* sensors. The former acquire range images, in which each pixel encodes the distance between the sensor and a point in the scene. The latter are the familiar TV cameras acquiring grey-level images. Our reference tasks, model acquisition and active inspection, form a logical sequence of relevance for applications:

ARTIFICIAL VISION
ISBN 0-12-444816-X
Copyright © 1997 Academic Press Ltd
All rights of reproduction in any form reserved

first, 3D models are acquired automatically, thus saving the tedious process of building them by hand with CAD packages; second, the models are used for inspection purposes, i.e. to determine automatically the best positions in space from which mobile sensors can inspect user-selected object features.

This chapter introduces the main ideas of model acquisition and active sensor planning. It contains a review of relevant literature, and presents examples of working systems for both model acquisition and sensor planning (Section 4.3). An introduction to range data and range sensors has also been included. The interested reader will find the extensive bibliography provided a useful starting point for more detailed investigations.

4.2 BACKGROUND

This section gives an overview of the literature of the three main areas touched upon: range data, model acquisition and active sensor planning.

4.2.1 Range image acquisition

Range sensors measure the distance between themselves and the objects in the world. As intensity images measure light intensity, range images measure distance, and are therefore a direct representation of *shape*. Range images are naturally rendered as 3D surfaces, showing the distance of each points from an arbitrary background (see example in Figure 4.1: the object portrayed is the casting of a mechanical component). Of course one can display a range image as a grey-level image, by associating different grey levels to different depths (see for instance Figures 4.3 and 4.4).

Range images are used in many applications, including bin picking, robotic assembly, inspection, recognition, robot navigation, automated cartography and medical diagnosis. This section introduces only the range sensors most frequently used in machine vision, namely non-contact devices based on *active optical principles*. The main interest of such devices consists in the ability of

Figure 4.1   Two views of a range image rendered as 3D surface.

collecting large amounts of accurate 3D coordinates in a reasonable time. Recent technical surveys on range sensors have been published by Besl (1988), Everett (Everett, 1989; Everett *et al.*, 1992) (both of which analyse the performance of several commercial and experimental sensors), Nitzan (1988) and Monchaud (1989).

4.2.1.1 *Optical range sensors*

The main advantage of range images over other types of images is that they represent *explicitly* the shape of surfaces. Basic range sensors can measure depth at *single points*, on *lines* (acquiring *range profiles*), or in 2D *fields of view* (acquiring *range images*). Multiple views are obtained by arranging multiple sensors in different locations, maneuvering one or more sensors around the scene with a robot, rotating and translating the scene itself, or adjusting a system of mirrors.

Besl (1988) reports six different optical principles on which optical, active range sensors are based: radar, triangulation, Moiré techniques, holographic interferometry, focusing and diffraction. The best absolute accuracies (0.1 nm) are obtained by *holographic interferometry*, but the very small maximum depth of field of this technique (about 0.1 mm) limits its applications to very specific domains.

Imaging radars can measure (a) the time of flight of pulses (sonar, laser) reflected by the surfaces in the scene; or (b) the phase difference between transmitted and received amplitude-modulated signals; or again (c) the beat frequency in a FM CW radar.

In *Moirè range-imaging sensors*, depth information is encoded in the phase difference component of an interference signal (pattern).

Active focusing infers range by analysing the blur in intensity images acquired across a range of focus values. This method has led to compact and inexpensive sensors, but with limited precision.

Diffraction sensors are based on the Talbot effect: if a periodic line grating is illuminated with coherent light, exact in-focus images of the grating are formed at regular periodic intervals (the ambiguity intervals of the sensor). Diffraction sensors can work at video rate but have not yet been investigated much.

Triangulation sensors are ostensibly the most popular range sensing method in computer vision and automatic model acquisition. In a typical sensor, a camera observes a pattern of structured light projected on the scene (e.g. a beam of laser light, or a grid of lines). If the geometry of the system is known (i.e. if the system has been calibrated), depth can be inferred by triangulation. Some triangulation devices provide registered range and intensity images (Monchaud, 1989). These sensors can reach accuracies of a few micrometers (over limited fields of view); the maximum field of view can be of several tens of meters. Triangulation systems can be built rather inexpensively and have therefore been very popular in academic research. Section 4.3.1 illustrates a typical triangulation, laser-based range sensor.

A special class of triangulation sensors are 3D *laser scanners*, or 3D *digitizers* (Wohlers, 1992), currently the main technology for surface digitization in industrial applications like reverse engineering of mechanical parts, creation of patterns for molded tooling and investment casting, digitization of anatomical parts for prostheses, implants and surgery, and creation of complex shapes and special effects for video and film production (3D scanners were used for the special effects of *The Abyss, Robocop 2, Terminator 2*). Most of the 3D laser scanners in use in Europe and North America are manufactured in the United States. Several manufacturers of 3D digitizers supply model editing packages to be used with their digitizers. These packages allow the user to fit CAD models to the data and to edit interactively both data and models. At present, most computer-generated models still require editing before being used, but the workload on the human operator is minor if compared to creating full CAD models from scratch.

Most existing 3D laser scanners can be classified as *single point* or *plane of light*. Single-point scanners capture one point at a time. The typical solution to read in a whole object is to rotate the object, thus generating a stack of evenly spaced contours. Plane-of-light systems project a light stripe on the object and read the depth of the points on the resulting 3D curve. They capture more quickly than single-point scanners, but typically measure points redundantly by producing overlapping grids of data. The resulting data file can be very large and require appropriate data reduction.

4.2.1.2 *Coordinate measuring machines*

Coordinate measuring machines are an alternative, well-established technology for digitizing shapes, or acquiring range images. They offer excellent accuracies over large working volumes. CMM measure shape by touching the tip of a probe to each surface point to measure. The measure is obtained by tracking the position of the probe as it moves in space: when the probe touches the surface, its 3D position is recorded. The probe can be mechanical, magnetic or optical. Probe motion is achieved either manually or by preprogramming the actuator on which the probe itself is mounted. The main disadvantage of CMM is that they involve complex programming for each new object, or skilled personnel to manually move direct the probe; acquiring enough points to describe complex surfaces can therefore be very slow, and depend heavily on operator guidance. Non-contact, optical sensors, on the contrary, require less mechanical motion than CMM, work independent of the shape to be measured, are faster, and do not risk destructive contact.

4.2.2 Automatic model acquisition

The literature of geometric model acquisition is ample, and we shall confine ourselves to a brief review of systems constructing 3D models from multiple range views. The review is organized by type of 3D representation generated (B-reps, volumetric models, triangulations, parallel contours).

4.2.2.1 *Acquiring B-rep models*

Various authors have concerned themselves with the problem of building boundary representations of solid objects, or *B-reps*. These describe an object in terms of its boundary, be it surfaces (patches) or contours (lines).

Potmesil (1983) proposed a method for constructing surface models of arbitrarily-shaped solid objects by matching the 3D surface segments describing these objects at different views. The surface models were composed of networks of bicubic parametric patches.

Koivunen *et al.* (1993) presented a system building CAD models from range data for CAD and manufacturing applications. Nonlinear filters based on robust estimation attenuate the noise in the data, which are interpreted by fitting models. The model construction strategy was based on the extraction of volumetric models (superellipsoids) from each segmented image, and on the use of the parameters of the volumetric description for selecting the appropriate CAD modeling primitive to be used (sphere, ellipsoid, parallelepiped, cylinder, cone, torus, wedge) to describe each part.

Koivunen and Vezien (Vezien and Koivunen, 1993; Koivunen *et al.*, 1993) reported model acquisition experiments using a registration algorithm based on the *iterative closest point* method (Besl and McKay, 1992) for integrating different 3D views of an object in one unique model.

Medioni's group at the University of Southern California reported research on model acquisition in several recent papers (Parvin and Medioni, 1991a; Parvin and Medioni, 1991b; Parvin and Medioni, 1992; Chen and Medioni, 1991; Chen and Medioni, 1992). The system sketched in Parvin and Medioni (1992) computes B-reps and curved objects bounded by quadric surfaces from multiple range images. Attributed graphs describing boundaries (first and second order discontinuities), surfaces, vertices and their relationships are extracted from each view (Parvin and Medioni, 1991b). Correspondence between graphs is obtained by minimizing a set of local, global and adjacency constraints.

The automatic generation of 3D polyhedral models from range data is approached by Connolly *et al.* (1987) by reconstructing a partially constrained solid model from each view and intersecting the models obtained from multiple views. Single-view models are built by extracting step and curvature (crease) edges from the data, then fitting planes to the resulting 3D grid of vertices and edges.

Vemuri and Aggarwal (Vemuri and Aggarwal, 1987; Vemuri and Aggarwal, 1988) discuss 3D model acquisition from multiple views in the framework of representation and recognition from dense range maps. The focus of their work is on matching descriptions from single unknown views to models and simple model acquisition techniques are adopted. Range data are acquired by a fixed sensor observing an object sitting on a turntable. The only parameter of the transformation is therefore a rotation angle, and this is easily determined by matching the object's silhouettes in different views. View integration is then performed simply by averaging data in overlapping regions. The final 3D model is expressed in terms of surface patches, whose shape is classified using the sign of the principal curvatures.

Sullivan *et al.* (1992) devised a method for the minimization of point-to-surface distances and use a combination of constrained optimization and nonlinear least-squares techniques to minimize the mean-squared geometric distance between a set of points or rays and a parametrized surface. In this way they can build 3D models by fitting algebraic surfaces to sets of 3D data (range and CT) as well as 2D image contours. The same paradigm is also used to estimate the motion and deformation of non-rigid objects in sequences of 3D images. Experimental examples are reported, including tracking the shape of the left ventricle of the beating heart of a dog over 16 CT images (the computing time for 600 points is about 15 minutes) and automatic model construction from a set of occluding contour points extracted from 5 to 7 views of an object (computing time with 300 contour points about 90 minutes).

Vemuri and Malladi (Vemuri and Malladi, 1991) proposed to use elastically deformable surfaces to acquire 3D models from one or more sets of depth data. An intrinsic parametrization of the surface was adopted, in which the parametric curves are the lines of curvature. The authors assumed that the transformation between views was known and used a cylindrical coordinate frame to assimilate the data.

Chen and Medioni (1991, 1992) described an algorithm for fusing different range images of an object. Their algorithm calculated accurately the registration between different views of an object, assuming knowledge of an initial approximation for the registration. This algorithm can be regarded as a specialization of Besl and MacKay's algorithm for the registration of free-form surfaces (Besl and McKay, 1992).

4.2.2.2   *Acquiring volumetric models*

Volumetric models describe objects by their main concentration of mass, e.g. a human body in terms of head, trunk, arms and legs. Of course, several levels of details are possible. Volumetric representations include generalized cylinders (Nevatia and Binford, 1977), octrees (Connolly *et al.*, 1987; Soucy *et al.*, 1992), superquadrics (Pentland and Sclaroff, 1991), and sticks/blobs/plates (Fitzgibbon, 1992).

Ferrie and Levine (1985) propose a full approach to characterize the shape of moving 3D objects. Their basis for object description is to compute physical attributes in terms of differential geometry at various levels of abstraction. The final models acquired are composed of ellipsoids and cylinders.

Pentland and Sclaroff (1991) present a closed-form solution for the problem of recovering 3D solid models from 3D surface measurements. The method is based on finite element analysis. An example of integration of multiple viewpoints is given, whereby three range views of a human figure are segmented into approximately convex volumes or 'blobs', and locally deformable models are fitted to the parts. The segmented data are related to the current recovered model by rotating the data. The transformation between the views is known. In another example, the authors demonstrate their fitting technique by recovering a CAD model of a human head using full 3D range data. An interesting feature of this work is that the representation used, called modal representation, is unique and can therefore be used for recognition. The authors mention two weaknesses of the technique: the need to estimate object orientation to $\pm 15°$, and the need to segment data into parts in a stable, viewpoint invariant manner.

Several researchers have investigated the automatic acquisition of volumetric models from *single views*. After the early work by Nevatia and Binford (1977) concentrating on generalized cylinders, systems capable of recovering generalized cones (Soroka, 1981), ellipsoids (Phillips and Rosenfeld, 1988), superquadrics (Pentland, 1990; Solina and Bajcsy, 1990; Ferrie *et al.*, 1990; Boult and Gross, 1987), suggestive models or SMS (Trucco, 1993; Fisher, 1992; Fisher *et al.*, 1994), have been reported.

4.2.2.3 *Acquiring polygonal approximations*

Much work has addressed the reconstruction of 3D *triangulations and polygonal approximations* of surfaces from multiple range views. Bhanu (1984) aimed at complex shapes, assuming the transformation between views known *a priori*. Different range views were obtained using a turntable which rotated the object around a fixed axis in space. A top and a bottom view, acquired from viewpoints belonging to the axis of the turntable, were registered with all the other views using three control points. The process acquired and integrated views sequentially. The final model was expressed by a polygonal approximation.

Soucy and others (Soucy and Laurendau, 1991; Soucy and Laurendau, 1992; Soucy *et al.*, 1992) describe a system which can integrate multiple range views of an object into a 3D model. The final model is a 3D Delaunay triangulation. Algorithms are described for computing optimal (equiangular) multi-resolution triangulations and for converting triangulated surfaces into octrees. Coarser surface models are obtained by a sequential optimization process which, at each iteration, removes the point that minimizes the approximation error in the local retriangulation. Fusion of multiple range views into a non-redundant surface model is done using a technique based on Venn diagrams. Intersections of range views are estimated by finding points featuring in both views.

4.2.2.4 *Acquiring parallel contours*

Parallel contours have also been used to represent automatically acquired models. They are a common choice for 3D scanners (Wohlers, 1992) when a small object can be rotated in front of the scanning head. A representative example is the work by Wang and Aggarwal (1989), who build 3D models made of parallel contours by combining active and passive sensing. A turntable allows acquiring several views of an object with known rotation angles. No correspondence is used and different views may not overlap. Contour-based systems are also reported by Phillips and Rosenfeld (1988) and Soroka (1981).

4.2.3 Sensor planning

Sensor planning is concerned with the automatic generation of strategies to determine configurations of sensor parameters that guarantee the optimal achievement of a sensing task. This implies that information about the sensors, the object and the task is available to the system. Such strategies embrace many parameters, including sensor placement in space (Sakane and Sato, 1991; Tarabanis *et al.*, 1994; Fitzgibbon and Fisher, 1992; Trucco *et al.*, 1992, 1994a), illumination placement (Sakane and Sato, 1991), intrinsic sensor parameters (Tarabanis *et al.*, 1994), and task-driven sensor selection (Ikeuchi and Robert, 1991). Three areas of research in sensor planning can be identified: *object feature detection*, *model-based recognition* and *scene reconstruction*. The systems in the first area (Cowan and Kovesi, 1988; Ikeuchi and Robert, 1991; Sakane and Sato, 1991; Tarabanis *et al.*, 1994) determine the values of the sensor parameters (e.g. position, orientation, optical settings) for which particular features of a known object in a known pose satisfy particular constraints when imaged. The systems in the second area are concerned with the selection of the most useful sensing operations for identifying an object or its pose (Kim *et al.*, 1985; Magee and Nathan, 1987; Ikeuchi and Robert, 1991). The third area is concerned with building a model of a scene incrementally, choosing new sensor configurations on the basis of the information currently available to the system about the world (Connolly, 1985; Maver and Bajcsy, 1990) and is therefore strictly related to model acquisition. The example implementation of Section 4.3.3 falls in the first area of this classification.

An essential element of a system predicting the aspects of an object from different viewpoints is its object representation. From this point of view we can identify two classes of 3D object representations: viewer-centered and object-centered representations. *Object-centered representations*, e.g.

generalized cylinders, octrees or superquadrics, assign a reference frame to each object and do not predict explicitly the object's appearance; in other words, they do not consider the observer as an element of the representation. *Viewer-centered representations*, on the contrary, model 3D objects through a number of *views*, obtained from different viewpoints in space. Views are defined by the relative position of object and sensor in space, and change therefore with the position of the observer. *Aspect graphs* (Bowyer and Dyer, 1990) are possibly the most popular form of viewer-centered representations. An *aspect* is a region of space associated with a set of object features visible simultaneously. When, moving the sensor or the object, some features appear or disappear (a situation called *visual event*), a new aspect is entered. The features considered in the literature are nearly invariably contours (Gigus *et al.*, 1988; Petitjean *et al.*, 1992; Eggert and Bowyer, 1991). Much discussion has taken place recently about the pros and cons of aspect graphs (Bowyer, 1991); the former being mainly the similarity of models and sensor data, which simplifies data-to-model comparisons; the latter their large size and the related indexing problems, their algorithmic complexity, and their mathematical nature, which often ignores that several aspects might never be detected by real sensors.

Most of the shortcomings of aspect graphs are circumvented by the use of *approximate visibility techniques*, commonly adopted in applications. Approximate representations (Fekete and Davis, 1984; Korn and Dyer, 1987; Silberberg *et al.*, 1984) restrict the set of possible viewpoints (the *visibility space*) to a sphere of large but finite radius, centered around the object. *A quasi-regular tessellation or geodesic dome* is then built on the sphere to cover its surface with a discrete grid of points. Each point is a viewpoint in the *approximate visibility space*. Feature visibility from each viewpoint is computed by raytracing CAD models of the objects. Approximate representations are a well-understood class of methods applicable to every object shape, and therefore very popular in applications. However, one cannot be certain that all important views are captured once fixed the *resolution* of the tessellation, i.e. the number of viewpoints.

4.3 EXAMPLES OF IMPLEMENTED SYSTEMS

4.3.1 A 3D range data sensor

This section illustrates the triangulation range sensor developed at the Artificial Intelligence Department of the University of Edinburgh (Trucco and Fisher, 1994; Trucco *et al.*, 1994b). The sensor architecture is sketched in Figure 4.2.

The object to be scanned sits on a platform moved on a linear rail by microstepper motors under computer (Sun3) control through a Compumotor CX interface with in-built RS-232C interface. One microstep is 6.6 μm, and the

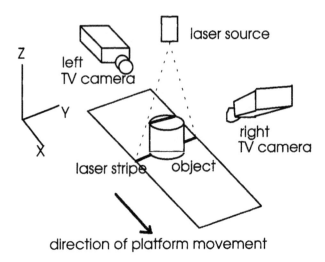

Figure 4.2   Architecture of the range sensor.

nominal positioning accuracy and repeatability of the motors is 2 microsteps. Objects must be contained in a parallelepipedal workspace about 15 cm each side.

The object is moved through a plane of laser light (output power 2 mW at 632.8 nm) obtained using a HeNe source, mirrors and a cylindrical lens. The laser beam has a circular cross-section with Gaussian intensity profile before passing through the cylindrical lens. The lens spreads the beam into a thin plane, so that the intensity profile in the workspace of the sensor is practically constant along the stripe and still Gaussian across the stripe. In practice, the cross-section of the stripe as observed by the cameras is not a perfect Gaussian, as each pixel integrates light over its field of view, the physical sensor pads of the solid-state cameras have gaps between them, the sensor pads have internal structure that affects their sensitivity, not all sensor pads are equally sensitive, and camera sensor and frame buffer can have different sizes.

The planar curve (stripe) resulting from the intersection of the plane of laser light with the object surface is observed by two opposing cameras (off-the-shelf 577 × 581 Panasonic BL200/B) mounted about one meter from the platform. This camera arrangement limits the occlusion problem and is very useful to enforce consistency constraints on the measurements (Trucco and Fisher, 1994). The acquired images are stored in a Datacube as 512 × 512 frames. In this setup, one millimeter in the vertical direction corresponds to less than one pixel in the images. Several parameters can be controlled by the user, including image scaling factors, depth quantization, image resolution, and the depth interval to be scanned.

4.3.2 A 3D model acquisition system

This section describes the 3D model acquisition system developed at the Department of Artificial Intelligence of the University of Edinburgh (Bispo *et al.*, 1993; Trucco and Fisher, 1992, 1995; Fisher *et al.*, 1994). Two essential issues are *model representation* and *view registration*. As seen in Section 4.2.2, several representations can be adopted to express the models, e.g. splines, surface patches, triangulations, volumetric descriptions and finite elements. The choice of a representation is linked intimately to the problem of estimating the transformation registering two successive views of the object.

The system described here uses two complementary model representations, each of which implies a different solution to the registration problem: a conventional, symbolic surface patch representation (Fan, 1990; Trucco and Fisher, 1992), illustrated in Figure 4.3 (left) is combined with a B-spline model (Figure 4.3 (right)). The symbolic model allows fast indexing from a large database, quick pose estimation (due to the small number of corresponding features), and easy specification of inspection plans (for example, the system can be instructed to 'measure diameter of hole 2 in patch B'). On the other hand, pose estimation is limited in accuracy by the small number of correspondences and errors in the surface fitting, and provides only an incomplete surface coverage: only the most stable surface patches are retained in the model. This lack of complete data is undesirable for reverse engineering tasks, and is cured by the use of spline models. Using these models, initial pose estimates can be optimized (albeit expensively), and complete surface models easily obtained.

The remainder of this section gives an overview of the computational modules of the system, namely: (a) constructing the surface patch representation and the spline model from each single range image (view); (b) estimating the transform between views by matching the surface patch models associated with two views; (c) refining the estimate using the spline model; and (d) merging the views and post-processing the current object model.

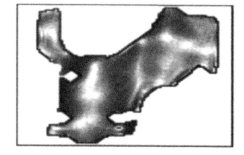

Figure 4.3   Example models produced by the system. Left: a surface patch model of the component in Figure 4.1 right: a coarse spline model.

4.3.2.1   *Surface representation from each range view*

Building the symbolic model   Objects are initially scanned using a laser triangulation system. Depth and orientation discontinuities are detected from the raw data, and used as *a priori* weights for diffusion smoothing (Trucco, 1992). Mean and Gaussian curvatures are calculated from the smoothed image, and the data are then divided into homogeneous regions of positive, negative or zero curvature. Each region is then approximated by a viewpoint-invariant biquadratic patch (Fitzgibbon and Fisher, 1993), and finally expanded to include neighbouring pixels which are within 3σ ($\sigma = 0.15$ mm) of the fitted surface. After this segmentation stage, the region boundaries are approximated by polylines and conics, and adjacent regions are intersected to produce more consistent boundaries. The resulting description (Figure 4.3 (right)) is converted into a vision-oriented modelling representation, called the Suggestive Modelling System or SMS (Fitzgibbon, 1992), for visualization and use in the model matching module.

Building the spline model.   The spline model is constructed by laying a regular grid of control points on the image and fitting a third-order B-spline to the data. Background and noise points are removed in advance. The density of the grid is currently determined by the required accuracy; a 50×50 sampling allows the object modelled in Figure 4.3 (right) to be approximated to within a maximum error of 0.8 mm. An obvious extension is to allocate the knot points more densely at regions of high curvature, as the curvature maps are available from the segmentation process.

4.3.2.2   *Estimating the transform between views*

We now wish to estimate the parameters of the rigid transformation which relates two views of an object, assuming the images overlap. We start by applying the segmentation process described above to each image, thus producing two lists of surface patches. From these, an interpretation tree algorithm (Grimson, 1990) finds consistent sets of corresponding pairs of surfaces. The pairs allows us to compute the 3D pose of the object using least-squares techniques. The pose is used as an initial estimate for an iterated extended Kalman filter (Waite *et al.*, 1993), which computes both the final pose estimate and its uncertainty.

The accuracy of view registration is within about 1σ of the noise on the range data if three or more linearly independent planar surfaces can be extracted reliably from the object. An example of automatic view registration is shown in Figure 4.4, in which the data of the new view are shown in light grey, and those of the matched model built from previous views in the dark grey. The image shows a good quality match (data and model are positioned very closely) using independent planar patches and patch centroids. If not enough planar surfaces are available for matching, biquadratic surfaces estimated about the patch

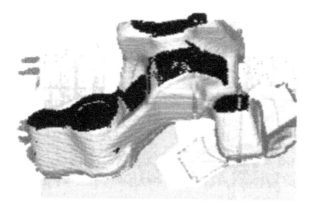

Figure 4.4 Surface-based matching results.

centroids are used to constrain the pose; in this case, translation accuracy falls to about 5 mm. If the pose needs to be constrained by using paired centroids, the system is open to error due to occlusion. The rotational accuracy of registration is generally within 1 degree.

4.3.2.3 *Refining the inter-view transform*

Given an initial pose estimate from the symbolic model matcher we can now use the spline model to refine the estimate. The pose is optimized using the Iterated Closest Point (ICP) algorithm (Besl and McKay, 1992). 2D experimental tests indicate that the region of convergence occupies about 25% of the space of all possible poses. In the 3D tests on the object above, a less complete investigation indicates convergence with up to 90 degrees of initial rotation error. The disadvantage of this technique is its computational complexity: for each data point, we must locate the nearest point on the model surface, then calculate the registering transform. Locating the closest point is sped up by a combination of multigridding and subsampling the basic gradient descent algorithm. The registration accuracy is 'optimal' in the sense that the noise statistics of the residuals are symmetric and white.

4.3.2.4 *View registration and model postprocessing*

Final processing on the models includes merging the single-view descriptions into a single reference frame. This is done easily thanks to the SMS representation for surface patches, which separates patch descriptions into shape, extent and position. The spline models may be treated similarly, by calculating a new fitting spline for the merged sets of range data.

4.3.3 An active inspection system

This section describes briefly a sensor planning system, GASP, which can plan optimal inspection actions for a variety of tasks, objects and sensors. We show various examples with an aerospace mechanical component in a simulated environment.

4.3.3.1 *Inspection scripts and inspection tasks*

Visual inspection strategies, that we shall call *inspection scripts*, depend on the type and number of both sensors and object features to be inspected: therefore, models of both sensors and objects must be available to the planner. At present, GASP can compute inspection scripts for both *single* or *multiple sensors*. A single sensor can be either an intensity or a range imaging camera. The only multiple sensor currently modelled in GASP is a stereo pair of intensity cameras (or *stereo head*). The inspection scripts that GASP can generate at present are the following:

- *Single-feature, single-sensor scripts* find the position in space from which a single imaging sensor (intensity or range) can inspect a single object feature optimally.
- *Single-feature, multiple-sensor scripts* find the position in space from which a stereo head can inspect a single object feature optimally.
- *Multiple-feature, single-sensor scripts* (a) find the position in space from which a single imaging sensor (intensity or range) can simultaneously inspect a set of features optimally; (b) find the best path in space taking a single imaging sensor (intensity or range) to inspect a set of object features from optimal positions.
- *Multiple-feature, multiple-sensor scripts* find the best path in space taking a stereo head to inspect a set of object features from optimal positions.

A special representation, the FIR (Feature Inspection Representation), makes inspection-oriented information explicit and easily accessible. In the next sections FIRs are used by GASP to create inspection scripts in a simulated environment, written in C on SPARC/Unix workstations. A FIR partitions the (discrete) set of all the viewpoints accessible to the sensor (the *visibility space*) into *visibility regions*, each formed by all the viewpoints from which a given feature is visible. Viewpoints are weighted by two coefficients, *visibility* and *reliability*. The former indicates the size of the feature in the image (the larger the image, the larger the coefficient); the latter expresses the expected confidence with which the feature can be detected or measured from a viewpoint by a given image processing module. For instance, estimates of the curvature of cylindrical surfaces from range images become seriously unreliable if the surface appears too small or too foreshortened (Trucco and Fisher, 1995).

Visibility and reliability are combined into an *optimality* coefficient, which quantifies the global merit of the viewpoint for inspecting the feature. The relative importance of visibility and reliability can be adjusted in the combination: it is therefore possible to bias GASP scripts according to task requirements. For instance, one might want to emphasize visibility when simply checking that an object feature is present; to ensure that a measurement is taken with maximum confidence, reliability should get a high weight.

In GASP, the visibility space is modelled by a *geodesic dome* centered on the object. The viewpoints are the centers of the dome's facets. The algorithm to compute the FIR generates a geodesic dome around a CAD model of the object to inspect, then raytraces the model from each viewpoint on the dome (Trucco *et al.*, 1992). Sensor models incorporated in the raytracer are used to determine the visibility and reliability for each viewpoint. FIR sizes vary with the number of object features, the resolution of the raytracer images, the density of the viewpoints on the dome, and the regions reachable by the sensors. A FIR for the widget shown in Figure 4.5 (left), observed from 320 viewpoints and with image resolution of 64×64 pixels, occupies about 80 kbytes. The widget is about 250 mm in length.

4.3.3.2   *Single-feature inspection*

In a FIR, the viewpoints belonging to a visibility region are arranged in a list ordered by optimality; therefore, the best viewpoint for inspecting a given feature is simply the first element of a list. Figure 4.5 (right) shows the visibility region (a portion of the geodesic dome representing the whole visibility space) of the planar patch highlighted in Figure 4.5 (left) given a single intensity camera of focal length about 50 mm and image resolution 64×64 pixels. The widget has been magnified for clarity. The object-camera distance (dome radius) was determined by GASP as the minimum one such that the whole widget is visible from all viewpoints; other choices are possible for the choice of the dome radius. Viewpoints have been shaded according to their optimality (the darker the

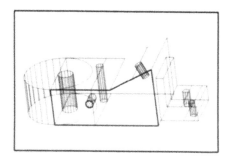

Figure 4.5   Left: CAD model of a widget and visibility region for top planar patch (highlighted).

better), assuming equal weights for visibility and reliability in the optimality computation.

4.3.3.3 Multiple-feature, single-sensor scripts

Case (a): simultaneous inspection of several features. In this case, a sensor must inspect several features simultaneously from one position in space. The region of the visibility space from which a set of features is simultaneously visible is called the *covisibility region*, and is obtained as the intersection of the visibility regions of the individual features. This is done simply and efficiently due to the FIR's structure. Notice that the merit (optimality) of a viewpoint for inspecting the set of features must be defined as a function of the optimality of the viewpoint for inspecting each *individual* feature (Trucco et al., 1992).

Case (b): sequential inspection of several features. Given a set of features, we now want to find a *path in space* which takes the sensor through the optimal inspection positions associated to all the features in the set; the path must also be as short as possible since we wish to minimize the number of sensor repositioning actions. Planning the shortest 3D path through a given set of points is a NP-complete problem, and only approximate solutions can be found. We considered three TSP algorithms: CCAO (Golden and Stuart, 1985), simulated annealing, and the elastic net method (Durbin and Willshaw, 1987), and chose the best one for our purposes by implementing them and testing their performances with a large number distributions of viewpoints to visit. We found that, with up to 100 viewpoints, CCAO outperformed the other two in terms of path length and distance from the overall optimal solution. CCAO was therefore incorporated in GASP. An example of inspection path is given in the section on multiple-sensor, multiple-feature scripts.

4.3.3.4 Multiple-sensor, single-feature scripts

At present, the only multiple-sensor configuration considered in GASP is a stereo camera pair, or stereo head. In this case, we want to determine the position in space from which the head can observe at best a given object feature. To do this, both cameras must be placed in good viewpoints for inspection, which is not trivial because good visibility for one camera does not necessarily imply the same for the other. Using a FIR and information about the head geometry, GASP selects efficiently the most promising admissible positions of the head inside the region, and picks the best one. We define the position of the stereo head as that of the midpoint of the line connecting the optical centers of the cameras (the head's *baseline*). Since the two cameras are placed in two distinct viewpoints, the optimality of the head position (*stereo optimality*) is defined as a combination of those of the viewpoints in which the two cameras are placed. Figure 4.6 (left) shows the optimal placement for a stereo head found

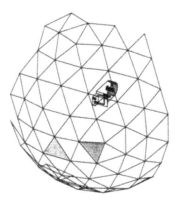 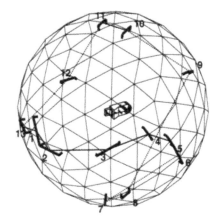

Figure 4.6 Left: stereo head inspection. Right: inspection path.

by GASP in the large (compared with the head's camera-to-camera distance) visibility region of one of the widget's back planar surfaces.

4.3.3.5 *Multiple-sensor, multiple-feature scripts*

The final class of inspection scripts find the shortest inspection path through the viewpoints from which each feature in a given set can be optimally inspected by a stereo head. Since in GASP the head's position is represented by a single point, it is possible to apply again the path-finding module used for the single-sensor case. An example of shortest 3D inspection path is shown in Figure 4.6 (right), which shows only the visible part of the dome. Solid lines connect points on visible dome facets, dotted lines lead to or from occluded viewpoints. The stereo head is represented by the baseline segment. The path was computed simulating two intensity cameras of focal length 50 mm and image resolution 64 × 64. The dome radius (object-camera distance) was then fixed by GASP at 2400 mm in order to ensure full visibility of the widget from all views. The head's baseline was 500 mm. In these conditions, 13 surfaces of the widget were visible to the cameras (others were too small or always mostly occluded, e.g. the bottom surfaces of holes), and each surface was associated to an optimal inspection viewpoint for the stereo head. The 3D path shown is the shortest one found by the CCAO algorithm through such viewpoints.

ACKNOWLEDGEMENTS

Section 4.3 summarizes work carried out (in addition to the author) by many people: E. Bispo, L.D. Cai, R.B. Fisher, A.W. Fitzgibbon, M.J.L. Orr and M. Waite at the Department of Artificial Intelligence, University of Edinburgh

(UK); M. Diprima at the University of Udine (Italy); M. Umasuthan at the Department of Computing and Electrical Engineering of Heriot-Watt University (Edinburgh, UK).

REFERENCES

Besl, P. (1988) Active optical imaging sensors. *Machine Vision and Applications*, 127–152.

Besl, P. and McKay, N. (1992) A method for registration of 3-D shapes. *IEEE Trans. Pattern Analysis Machine Intelligence* **PAMI-14,** 2, 239–256.

Bhanu, B. (1984) Representation and shape matching of 3-D objects, *IEEE Trans. Patt. Anal. Mach. Intell.* **PAMI-6,** 3, 341–351.

Bispo, E., Fitzgibbon, A.W. and Fisher, R.B. (1993) Visually salient reconstruction of 3-D models from range data. *Proc. British Machine Vision Conf.*, Surrey, UK.

Blake, A. and Yuille, A. (eds) (1993) *Active vision.* MIT Press, Cambridge, Massachusetts.

Boult, T. and Gross, A. (1987) Recovery of superquadrics from depth information. *Proc. AAAI workshop on spatial reasoning and multi-sensor integration*, 128–137.

Bowyer, K. and Dyer, C.R. (1990) Aspect graphs: an introduction and survey of recent results. *Int. Journal of Imaging Systems and Technology* **2,** 315–328.

Bowyer, K. (ed.) (1991) Why aspect graphs are not (yet) practical for computer vision, panel discussion. *Proc. of IEEE workshop on advances in CAD-based vision*, Hawaii.

Chen, Y. and Medioni, G. (1991) Object modeling by registration of multiple range images. *Proc. IEEE Conf. on Computer Vision and Pattern Recognition* (CVPR-91), 2724–2729.

Chen, Y. and Medioni, G. (1992) Object modeling by registration of multiple range images. *Image and Vision Computing* **10,** 3, 145–155.

Connolly, C.I. (1985) The determination of the next best view. *Proc. IEEE Conf. on Robotics and Automation*, 432–435.

Connolly, C.I., Mundy, J., Stenstrom, J. and Thompson, D. (1987) Matching from 3-D range models into 2-D intensity scenes. *Proc. Int. Conf. on Comp. Vision* (ICCV-87), 65–71.

Cowan, C.K. and Kovesi, P.D. (1988) Automatic sensor placement for vision task requirements. *IEEE Trans. Patt. Anal. Mach. Intell.* **PAMI-10,** 407–416.

Durbin, R. and Willshaw, D. (1987) The travelling salesman problem. *Nature*, 689–691.

Eggert, D. and Bowyer, K. (1991) Perspective projection aspect graphs of solids of revolution: an implementation. *Proc. 7th Scandinavian Conf. on Image Analysis*, 299–306.

Everett, H.R. (1989) Survey of collision avoidance and ranging sensors for mobile robots. *Robotics and Autonomous Systems*, **5,** 5–67.

Everett, H.R., DeMuth, D.E. and Stitz, E.H. (1992) Survey of collision avoidance and ranging sensors for mobile robots. Revision 1, Tech. Rep. 1194, Naval Ocean Systems Center, San Diego, California.

Fan, T.-J. (1990) *Describing and recognising 3-D objects using surface properties.* Springer-Verlag, Berlin, Germany.

Fekete, G. and Davis, L.S. (1984) Property spheres: a new representation for 3-D object recognition. *Proc. IEEE Workshop on Computer Vision, Representation and Control*, 192–201.

Ferrie, F. and Levine, M. (1985) Piecing together the shape of 3-D moving objects: an overview. *Proc. IEEE Conf. on Comp. Vision and Patt. Rec.* (CVPR-85), 574–583.

Ferrie, F., Lagarde, J. and Waite, P. (1990) Recovery of volumetric object descriptions from laser rangefinder images. *Proc. 1st Eur. Conf. on Comp. Vision* (ECCV-90).

Fisher, R.B. (1992) Representation, extraction and recognition with 2nd-order topographic surface features. *Image and Vision Computing* 10, 3, 156–169.

Fisher, R.B., Fitzgibbon, A.W., Waite, M., Trucco, E. and Orr, M.J.L. (1994) Recognition of complex 3-D objects from range data. In: *Progress in image analysis and processing III* (ed. S. Impedovo), 599–606. World Scientific, Singapore.

Fitzgibbon, A.W. and Fisher, R.B. (1992) Practical aspect graph derivation incorporating feature segmentation performance. *Proc. British Machine Vision Conf.*, 580–589, Leeds, UK.

Fitzgibbon, A.W. (1992) Suggestive modelling for machine vision. *Proc. SPIE-1830 Int. Conf. on Curves and Surfaces in Computer Vision and Graphics III*, OE/Technology '92, Boston, Massachusetts.

Fitzgibbon, A.W. and Fisher, R.B. (1993) Invariant fitting of arbitrary single-extremum surfaces. *Proc. Machine Vision Conf.*, 569–578.

Gigus, Z., Canny, J. and Seidel (1988) Efficiently computing the aspect graph of polyhedral objects. *Proc. IEEE Int. Conf. on Computer Vision*, 30–39.

Golden, B.L. and Stewart, R. (1985) Empirical analysis and heuristics. In: *The travelling salesman problem* (eds Lawler, Lenstra, Rinnoykan and Shmoys). Wiley, New York.

Grimson, W.E.L. (1990) *Object recognition by computer: the role of geometric constraints.* MIT Press, Cambridge, Massachusetts.

Ikeuchi, K. and Robert, J.-C. (1991) Modeling sensor detectability with the VANTAGE geometric/sensor modeler. *IEEE Trans. on Robotics and Automation* RA-7, 771–784.

Kim, H.-S., Jain, R.C. and Volz, R.A. (1985) Object recognition using multiple views. *Proc. IEEE Conf. on Robotics and Automation*, 28–33.

Koivunen, V., Vezien, J.M. and Bajcsy, R. (1993) Procedural CAD models from range data. *Towards World Class Manufacturing '93*, Phoenix, Arizona.

Korn, M.R. and Dyer, C.R. (1987) 3-D multiview object representations for model-based object recognition. *Pattern Recognition* 20, 91–103.

Magee, M. and Nathan, M. (1987) Spatial reasoning, sensor repositioning, and disambiguating in 3-D model based recognition. *Proc. AAAI Workshop on Spatial Reasoning and Multi-Sensor Fusion*, 262–271. Morgan Kaufmann, Los Altos, California.

Maver, J. and Bajcsy, R. (1990) How to decide from the first view where to look next. *Proc. Image Understanding Workshop*, 482–496.

Monchaud, S. (1989) 3-D Vision and range finding techniques. *Optics and Lasers in Engineering* 10, 161–178.

Nevatia, R. and Binford, T. (1977) Description and recognition of curved objects. *Artificial Intelligence* 38, 77–98.

Nitzan, D. (1988) Three dimensional vision structure for robotic applications. *IEEE Trans. Patt. Anal. Mach. Intell.* PAMI-10, 3, 291–309.

Parvin, B. and Medioni, G. (1991a) A layered network for correspondence of 3-D objects. *Proc. IEEE Int. Conf. on Robotics and Automation*, 1808–1813.

Parvin, B. and Medioni, G. (1991b) A dynamic system for object description and correspondence. *Proc. IEEE Conf. on Comp. Vision and Patt. Rec.* (CVPR-91), 1602–1607.

Parvin, B. and Medioni, G. (1992) B-reps from unregistered multiple range images. *Proc. IEEE Int. Conf. on Robotics and Automation*, 1602–1607.

Pentland, A. (1990) Automatic extraction of deformable part models. *Int. Journ. Comp. Vision* 4, 107–126.

Petitjean, S., Ponce, J. and Kriegman, D.J. (1992) Computing exact aspect graphs of curved objects: algebraic surfaces. *Int. Journ. Comp. Vision* 9.

Phillips, T. and Rosenfeld, A. (1988) Decomposition of 3-D objects into compact subobjects by analysis of cross-sections. *Image and Vision Computing* **6**, 33–51.

Potmesil, M. (1983) Generating models of solid objects by matching 3-D surface segments. *Proc. Int. Joint Conf. on Artificial Intelligence* (IJCAI-83), 1089–1093.

Sakane, S. and Sato, T. (1991) Automatic planning of light source and camera placement for an active photometric stereo system. *Proc. IEEE Int. Conf. on Robotics and Automation*, 1080–1087.

Pentland, A. and Sclaroff, S. (1991) Closed form solution for physically based shape modeling and recovery. *IEEE Trans. Patt. Anal. Mach. Intell.* **PAMI-13**, 7, 715–729.

Silberberg, T.M., Davis, L. and Harwood, D. (1984) An iterative Hough procedure for three-dimensional object recognition. *Pattern Recognition* **17**, 621–629.

Solina, F. and Bajcsy, R. (1990) Recovery of parametric models from range images. *IEEE Trans. Patt. Anal. Mach. Intell.* **PAMI-12**, 2, 131–147.

Soucy, M. and Laurendau, D. (1991) Building a surface model of an object using multiple range views. *Proc. SPIE Conf. on Intell. Robots and Comp. Vision X: Neural, Biological and 3-D Methods.*

Soucy, M. and Laurendau, D. (1992) Multi-resolution surface modeling from multiple range views. *Proc. IEEE Conf. on Comp. Vision and Pattern Recognition* (CVPR-92).

Soucy, M., Croteau, A. and Laurendau, D. (1992) A multi-resolution surface model for compact representation of range images. *Proc. IEEE Int. Conf. on Robotics and Automation*, 1701–1706.

Soroka, B. (1981) Generalized cones from serial sections. *Computer Graphics and Image Processing* **15**, 154–166.

Sullivan, S., Ponce, J. and Sandford, L. (1992) On using geometric distance fits to estimate 3-D object shape, pose, and deformation from range, CT and video images. Technical Report, Beckman Institute, University of Illinois.

Tarabanis, K., Tsai, R.Y. and Allen, P.K. (1994) Analytical characterisation of the feature detectability constraints of resolution, focus, and field-of-view for vision sensor planning. *CVGIP: Image Understanding* **59**(3), 340–358.

Trucco, E. (1992) On shape-preserving boundary conditions for diffusion smoothing. *Proc. IEEE Int. Conf. on Robotics and Automation*, Nice, France.

Trucco, E. (1993) Part segmentation of slice data using regularity. *Signal Processing*, **32**, 1–2, Special issue on intelligent systems for signal and image understanding (ed. V. Roberto), 73–90.

Trucco, E. and Fisher, R.B. (1992) Computing surface-based representations from range images. *Proc. IEEE Int. Symposium on Intelligent Control* (ISIC-92), Glasgow, UK.

Trucco, E. and Fisher, R.B. (1994) Acquisition of consistent range data using local calibration. *Proc. IEEE Int. Conf. on Robotics and Automation*, 3410–3415, San Diego, California.

Trucco, E. and Fisher, R.B. (1995) Experiments in curvature-based segmentation of range data. *IEEE Trans. Patt. Anal. Mach. Int.* **PAMI** **17**(2), 177–181.

Trucco, E., Thirion, M., Umasuthan, M. and Wallace, A.M. (1992) Visibility scripts for active feature-based inspection. *Proc. British Machine Vision Conf.*, Leeds, UK.

Trucco, E., Diprima, M. and Roberto, V. (1994a) Visibility scripts for active feature-based inspection. *Patt. Rec. Letters* **15**, 1151–1164.

Trucco, E., Fisher, R.B. and Fitzgibbon, A.W. (1994b) Direct calibration and data consistency in 3-D laser scanning. *Proc. British Machine Vision Conf.*, York, UK.

Vemuri, B. and Aggarwal, J. (1987) Representation and recognition of objects from dense range maps. *IEEE Trans. Circuits and Systems* **CAS-34**, 11, 1351–1363.

Vemuri, B. and Aggarwal, J. (1988) 3-D model construction from multiple views using range and intensity data. *Proc. IEEE Conf. on Comp. Vision and Patt. Rec.* (CVPR-88), 435–437.

Vemuri, B. and Malladi, R. (1991) Deformable models: canonical parameters for surface representation and multiple view integration. *Proc. IEEE Conf. on Comp. Vision and Patt. Rec.* (CVPR-91), 724–725.

Vezien, J.M. and Koivunen, V. (1993) Registration and data description in geometric model construction from range data. *Proceedings of SPIE 2067*, Boston, Massachusetts.

Waite, M., Orr, M.J.L. and Fisher, R.B. (1993) Statistical partial constraints for 3-D model matching and pose estimation problems. *Proc. British Machine Vision Conf.*, Surrey, UK.

Wang, Y. and Aggarwal, J. (1989) Integration of active and passive techniques for representing 3-D objects. *IEEE Trans. Robotics and Automation* **RA-5,** 4, 460–471.

Wohlers, T. (1992) 3-D digitisers. *Computer Graphics World*, 73–77.

5

The Regularization of Early Vision

5.1 INTRODUCTION

A basic tenet of many computational studies of vision is the distinction between early vision and higher-level processing. The term *early vision* denotes those problems (like edge detection, motion estimation, depth recovery, and shape from X) that appear to be addressed at the first few stages of the processing of visual information. In both natural and machine vision systems the understanding of early vision is considered a necessary step for the accomplishment of higher-level visual tasks like manipulation, object recognition and scene description.

Unlike higher-level vision, early vision seems to be constituted by a number of roughly independent visual modules. Each module, at least to a first

ARTIFICIAL VISION
ISBN 0-12-444816-X

Copyright © 1997 Academic Press Ltd
All rights of reproduction in any form reserved

approximation, can be regarded as a bottom-up process characterized by a strong geometric flavor.

In this chapter a framework for the analysis and solution to early vision problems is reviewed. The framework, originally proposed by (Bertero *et al.*, 1988) is based on the theory of regularization of ill-posed problems (Tikhonov and Arsenin, 1977). An ill-posed problem (Hadamard, 1902; Courant and Hilbert, 1962) is a problem for which at least one of the conditions of *uniqueness, existence* or *continuous dependence* of the solution *on the data* is not ensured. The theory of regularization provides mathematical techniques able to restore these conditions for the solution to an ill-posed problem. The relevance of regularization theory to early vision, as pointed out by (Poggio *et al.*, 1985), is due to the fact that many early vision problems appear to be ill-posed.

The rest of this chapter is organized as follows. Section 5.2 describes the basic geometry and photometry of a visual system and formulates a few early vision problems as inverse problems. In Section 5.3 the mathematics of ill-posed problems and regularization theory is reviewed. In Section 5.4, the regularization of the early vision problems of Section 5.2 is outlined. The merits and limitations of this approach to computer vision are discussed in Section 5.5. Finally, Section 5.6 summarizes the presented results.

5.2 BASIC CONCEPTS

In this preliminary section the inverse (and ill-posed) nature of early vision problems is illustrated. To this purpose some basic notation of geometry and photometry is first established.

5.2.1 Preliminaries

In what follows a simple camera model is described and the fundamental equation of image formation is recalled. For a more detailed discussion of the geometry of vision the interested reader should refer to the appendix of the book by Mundy and Zisserman (1992), while for a comprehensive description of the physics of image formation the classic book by Born and Wolf (1959) is recommended.

5.2.1.1 *Geometry*

In the pinhole model of Figure 5.1 the viewing camera consists of a plane π, the image plane, and a point O, the focus of projection. Let $f > 0$ be the focal distance (i.e., the distance between π and O). In the 3D Cartesian system of

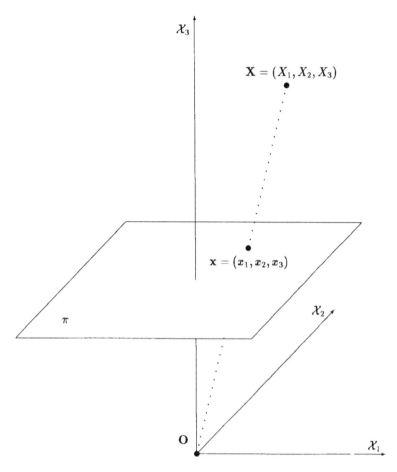

Figure 5.1   Pinhole camera model.

coordinates (χ_1, χ_2, χ_3) of Figure 5.1, the point $x = (x_1, x_2, x_3) \in \pi$, image of
the point $X = (X_1, X_2, X_3)$ in 3D space, is obtained through the perspective
projection of X on the image plane π with respect to the focus O. As shown in
Figure 5.1, x is the point in which the straight line through X and O impinges the
image plane π.

If the image plane π is the set of points $x = (x_1, x_2, x_3)$ with $x_3 = f$, the
fundamental equation of perspective projection can be written

$$x_i = f \frac{X_i}{X_3}, \qquad i = 1, 2, 3. \tag{5.1}$$

If the point X in 3D space is moving with velocity $V = (V_1, V_2, V_3)$, then
$v = (v_1, v_2, v_3)$, the velocity of the corresponding image point x, can be obtained

by differentiating eqn (5.1) with respect to time, that is,

$$v_i = f \frac{X_3 V_i - X_i V_3}{X_3^2}, \qquad i = 1, 2, 3. \tag{5.2}$$

Since $v_3 = 0$ for all the points of π, the vector v, also called the *motion field*, can be regarded as a 2D vector field $v = (v_1, v_2)$. Let us now recall some basic definitions of photometry.

5.2.1.2 *Image formation*

The *image irradiance* $E = E(x)$ is the power per unit area of light at each point x of the image plane. The *scene radiance* $L = L(X, \theta)$ is the power per unit area of light that can be thought of as emitted by each point X of a surface in 3D space in the particular direction θ. This surface can be fictitious, or it may be the actual radiating surface of a light source, or the illuminated surface of a solid.

In the pinhole camera model, the camera has an infinitesimally small aperture and, hence, the image irradiance at x is proportional to the scene radiance at the point X in the direction θ_0 (the direction of the line through X and O). In practice, however, the aperture of any real optical device is finite and not very small (ultimately to avoid diffraction effects). Assuming that (i) the surface is Lambertian, i.e. $L = L(X)$, (ii) there are no losses within the system and (iii) the angular aperture (on the image side) is small, it can be proved (Born and Wolf, 1959) that

$$E(x) = L(X) \omega \cos^4 \phi$$

where ω is the solid angle corresponding to the angular aperture and ϕ the angle between the principal ray (that is the ray passing through the center of the aperture) and the optical axis. With the further assumption that the aperture is much smaller than the distance of the viewed surface, the Lambertian hypothesis can be relaxed to give (Horn and Sjoberg, 1979)

$$E(x) = L(X, \theta_0) \omega \cos^4 \phi \tag{5.3}$$

with θ_0 direction of the principal ray. In what follows it is assumed that the optical system has been calibrated so that eqn (5.3) can be rewritten as

$$E(x) = L(X, \theta_0). \tag{5.4}$$

Equation (5.4), known as the image irradiance equation, says that the image irradiance at x equals the scene radiance at X in the direction identified by the line through X and O.

We are now in a position to formulate some of the classical problems of early vision as inverse problems.

5.2.3 Early vision as inverse optics

As already mentioned, the aim of early vision is the reconstruction of the physical properties of surfaces in 3D space from 2D images. This reconstruction amounts to finding the solution to different problems of inverse optics. Let us illustrate this point further by means of three specific examples.

5.2.2.1 *Edge detection*

The first step of many visual algorithms can be thought of as the problem of inferring the discontinuities of surface orientation, depth, texture and other physical properties of the viewed surfaces in 3D space from the sharp changes, or *edges* of the image brightness pattern (Shanmugam *et al.*, 1965; Canny, 1983; Torre and Poggio, 1986).

In essence, edge detection requires the preliminary computation of spatial derivatives of the image brightness. As it can easily be seen in the 1D case, the inverse (and ill-posed) nature of edge detection resides in the differentiation stage. Let $g = g(x)$ be a function defined on the interval $[0, 1]$. The derivative $g' = u$ of g can be computed as the solution to the equation

$$g(x) = \int_0^x u(t)\, dt. \tag{5.5}$$

Clearly, the problem of determining $u(x)$ from eqn (5.5) is an inverse problem. The problem is also ill-posed since the derivative u does not depend continuously on g. This can be best seen by means of a simple example. If

$$\hat{g} = g + \epsilon \sin \beta x,$$

then for all $x \in [0, 1]$ and independently of β

$$|\hat{g} - g| \leqslant \epsilon. \tag{5.6}$$

However, if $u = g'$ and $\hat{u} = \hat{g}'$, then

$$\hat{u} = u + \epsilon \beta \cos \beta x$$

and for $\beta \geqslant N/\epsilon$ and some x

$$|\hat{u} - u| \geqslant N. \tag{5.7}$$

From eqns (5.6) and (5.7) it follows that the derivatives of two arbitrarily close functions can be arbitrarily different.

Let us now discuss the problem of the computation of optical flow.

5.2.2.2 *Computing optical flow*

The estimation of the motion field v of eqn (5.2) from a sequence of images (Horn and Schunck, 1981; Hildreth, 1984; Adelson and Bergen, 1985; Heeger,

1987) is an essential step for the solution to problems like 3D motion and structure reconstruction (Longuet-Higgins and Prazdny, 1981; Rieger and Lawton, 1985; Heeger and Jepson, 1992).

It has been proposed to compute v from the apparent motion of the image brightness pattern $E = E(x, t)$ over time. For the sake of simplicity let us assume that this apparent motion, usually called *optical flow*, coincides with the motion field $v = (v_1, v_2)$ (for a critical discussion of the conditions under which optical flow and motion field are the same, see Verri and Poggio (1989)).

Probably the most common hypothesis on the changing image brightness is that the total derivative of E vanishes identically (Horn and Schunck, 1981), or

$$\frac{\mathrm{d}E(x, t)}{\mathrm{d}t} = 0. \tag{5.8}$$

The explicit evaluation of the left hand side of eqn (5.8) reads

$$E_{x_1} v_1 + E_{x_2} v_2 = -E_t, \tag{5.9}$$

where E_{x_1}, E_{x_2} and E_t are the partial derivatives of $E = E(x, t)$ with respect to x_1, x_2 and t respectively. From eqn (5.9) the inverse nature of the problem of computing optical flow from the spatial and temporal derivatives of the image brightness is apparent. Clearly, the problem is also ill-posed since the solution to eqn (5.9) (one equation for two unknowns, v_1 and v_2) is not unique.

5.2.2.3   Shape from shading

The visual information that can be extracted from the shading of an object can be usefully employed for the recovery of the shape of the viewed surface (Horn, 1975). Following Horn (1990) let us derive the fundamental equation of *shape from shading*.

First, it is assumed that the depth range is small compared with the distance of the scene from the viewer. Consequently, the projection is approximately orthographic and by means of a suitable rescaling of the image coordinates eqn (5.1) can be rewritten

$$x_i = X_i, \qquad i = 1, 2. \tag{5.10}$$

Now, if (i) the reflective properties of the viewed surface are uniform, (ii) there is a single, far away light source (that is, the incident direction is constant) and (iii) the viewer is also far away (that is, the viewer direction is approximately the same for all the points of the scene), then the scene radiance depends on the orientation and not on the position of a surface patch in the scene.

Therefore, by means of eqn (5.10) the scene radiance L at the point X in the direction θ_0 (the same direction for every X from (iii)) can be written in terms of the spatial derivatives of $f(x)$, the 'height' function which describes the viewed surface in terms of the image coordinates, or

$$L(X, \theta_0) = R(p(x), q(x)) \tag{5.11}$$

with

$$p = f_{x_1} \text{ and } q = f_{x_2}$$

partial derivatives of $f = f(x)$ with respect to x_1 and x_2 respectively and components of $\text{grad} f$, the gradient of f. The function $R(p,q)$, which is the scene radiance as a function of the components of $\text{grad} f$, is called the *reflectance map* (Horn and Sjoberg, 1979). Through eqn (5.11), the image irradiance equation (5.4) reads

$$E(x) = R(p(x), q(x)). \tag{5.12}$$

The recovery of shape from eqn (5.12), the fundamental equation of shape from shading, is clearly an inverse (nonlinear) problem. The problem is also ill-posed since eqn (5.12) is only one equation for two unknowns, p and q.

Let us now briefly review the fundamental concepts of regularization theory.

5.3 ILL-POSED PROBLEMS AND REGULARIZATION THEORY

In this section the distinction between well-posed and ill-posed problems is made precise. Then, the main techniques for determining the solution to (linear) ill-posed problems are briefly reviewed. For an exhaustive treatment of this subject see Tikhonov and Arsenin (1977) and Bertero (1989).

5.3.1 Definitions

For the sake of simplicity the presented analysis is restricted to the particular case of linear, inverse problems in functional spaces. With the only exception of *shape from shading* this assumption is sufficient to our purpose. Let X and Y be Hilbert spaces and L a continuous, linear operator from X to Y. The problem of determining $u \in X$ such that

$$g = Lu \tag{5.13}$$

for some $g \in Y$ is an inverse problem (the solution to eqn (5.13) requires the inversion of the operator L). We have the following definition:

The problem of solving eqn (5.13) is well posed (Courant and Hilbert, 1962) if the following conditions are satisfied:
1) for each $g \in Y$ the solution u is unique (uniqueness);
2) for each $g \in Y$, there exists a solution $u \in X$ (existence);
3) the solution u depends continuously on g (continuity).
Otherwise, the problem is said to be ill posed.
Now, let

$$N(L) = \{ f \in X, \text{ with } Lf = 0 \}$$

be the null space of L, or the set of the invisible objects, and

$$R(L) = \{g \in Y, \text{ with } g = Lf \text{ and } f \in X\}$$

the range of L, or the set into which the operator L maps X.

From the definition of null space of L, it follows that if $N(L) = \{0\}$, then L is injective and the solution is always unique. Similarly, from the definition of range of L if $Y = R(L)$, then L is *onto* and the solution to eqn (5.13) exists for all the $g \in Y$. Since L is linear, condition 3 (continuous dependence on the data) is implied by conditions 1 and 2. Therefore, if L is injective and *onto*, the problem of solving eqn (5.13) is always well posed.

Let us now briefly survey the main techniques that can be employed for the solution to eqn (5.13) when $R(L) \neq Y$ and $N(L) \neq \{0\}$.

5.3.2 Generalized inverse

If $R(L) \neq Y$ and the datum g does not belong to $R(L)$ the existence condition is clearly violated and the solution to eqn (5.13) does not exist. A least square solution or *pseudosolution* can then be defined as the function $u \in X$ such that

$$\|Lu - g\| = \inf\{\|Lf - g\|, f \in X\}$$

It can easily be seen that a pseudosolution exists if and only if $R(L)$ is closed. Furthermore, the pseudosolution is unique if and only if $N(L) = \{0\}$.

If $N(L) \neq \{0\}$ but $R(L)$ is closed, a pseudosolution u^+ of minimum norm, called a *normal pseudosolution* or *generalized solution*, can be found. It can be shown that the generalized solution u^+ is unique and orthogonal to $N(L)$. Since the mapping $g \to u^+$ is continuous, it follows that if $R(L)$ is closed the problem of determining u^+ from g is always well-posed. The operator L^+ defined by the equation

$$L^+ g = u^+ \tag{5.14}$$

is called the *generalized inverse* of L.

5.3.3 Condition number

It is important to notice that condition 3 (continuous dependence on the data) is weaker than stability or robustness of the solution against noise. The solution to a well-posed problem can still be severely ill conditioned. Let us discuss this point further in the assumption that $R(L)$ is closed.

From eqn (5.14) and the linearity of L we have

$$\|\Delta u^+\| \leq \|L^+\| \|\Delta g\|. \tag{5.15}$$

Similarly, from eqn (5.13) with $u = u^+$ it follows that

$$\|g\| \leq \|L\| \|u^+\|. \tag{5.16}$$

By combining eqns (5.15) and (5.16) it is easy to obtain

$$\frac{\|\Delta u^+\|}{\|u^+\|} \leqslant \alpha \frac{\|\Delta g\|}{\|g\|}$$

with $\alpha = \|L\| \|L^+\| \geqslant 1$. The *condition number* α controls the stability of the generalized inverse. If α is not much larger than 1, the problem of solving eqn (5.13) through the computation of the generalized inverse is said to be *well conditioned*. Intuitively, small perturbations of the data of a well-conditioned problem cannot produce large changes in the solution. Instead, if $\alpha \gg 1$, small perturbations may cause large changes and the problem is *ill conditioned*.

Let us now turn to the more difficult (and interesting) case in which $R(L)$ is not closed.

5.3.4   Regularized solutions

Regularization theory (Tikhonov and Arsenin, 1977) was developed in the attempt to provide a solution to ill-posed problems in the general case. The devised regularization methods have two advantages over the theory of generalized inverse. First, the methods can be applied independently of the closedness of the range of L. Second, the numerical solution to a regularized problem is well conditioned. Consequently, *regularized* solutions are also stable against quantization and noise.

The key idea of regularization is that in order to avoid wild oscillations in the solution to an ill-posed problem, the approximate solution to eqn (5.13) has to satisfy some smoothness (i.e., regularizing) constraint. Following the original formulation by Tikhonov let C be a constraint operator defined by

$$Cu = \sum_{r=0}^{R} \int c_r(x)|u^{(r)}(x)|^2 \, dx$$

where the weights c_r are strictly positive functions and $u^{(r)}$ denotes the rth order derivative of u. A typical regularization method consists of minimizing the functional

$$\Phi_\lambda[u] = \|Lu - g\|^2 + \lambda \|Cu\|^2. \tag{5.17}$$

with $\lambda > 0$. Depending on the available *a priori* information on the solution, the parameter λ can be determined in three different ways.

1)   If the approximate solution u is known to satisfy the constraint

$$\|Cu\| \leqslant E,$$

then the problem is to find the function u that minimizes the functional

$$\|Lu - g\|$$

with $\|Cu\| \leqslant E$. By means of the method of Lagrange multipliers the solution to this problem is equivalent to determining the minimum u_λ of Φ_λ with arbitrary λ, and to the search of the unique λ such that

$$\|Cu_\lambda\| = E.$$

2) If the approximate solution u is known to satisfy the constraint

$$\|Lu - g\| \leqslant \epsilon,$$

then the problem is to find the function u that minimizes the functional

$$\|Cu\|$$

with $\|Lu - g\| \leqslant \epsilon$. By means of the method of Lagrange multipliers the solution to this problem is equivalent to determining the minimum u_λ of Φ_λ with arbitrary λ, and to the search of the unique λ such that

$$\|Lu_\lambda - g\| = E.$$

3) If the approximate solution u is known to satisfy both the constraints

$$\|Cu\| \leqslant E \text{ and } \|Lu - g\| \leqslant \epsilon,$$

then the problem reduces to determining the minimum of Φ_λ with

$$\lambda = \left(\frac{\epsilon}{E}\right)^2.$$

The first method looks for the function that best approximates the data in the set of sufficiently regular functions, the second method for the most regular function in the set of functions sufficiently close to the data. In the third method, a compromise between the degree of regularity and the closeness of the solution to the data is established.

Let us conclude this very brief overview of regularization theory with three observations. First, if L is a convolution operator (as in the case of eqn (5.5)) the regularized solution can be simply obtained as a 'filtered' version of the non-regularized solution (Tikhonov and Arsenin, 1977).

Second, let us mention the existence of methods (like cross validation (Wahba, 1977) and generalized cross validation (Craven and Wahba, 1979) that can be usefully employed for the determination of the parameter λ in the absence of reliable *a priori* information on the degree of regularity and closeness of the solution to the data.

Third, if the operator L is nonlinear, regularized solutions can still be obtained by minimizing the functional Φ_λ of eqn (5.17) (with L a nonlinear operator). Under rather broad conditions on L and C, the existence and continuous dependence of the solution on the data can easily be proved (Tikhonov and Arsenin, 1977). In general, the uniqueness of the regularized solution is an open problem.

5.4 EARLY VISION REVISITED

In this section, the regularization of the problems of early vision formulated in Section 5.2 is outlined.

5.4.1 Edge detection

The typical procedure for the regularization of the problem of edge detection consists of two steps. In the first step the data are approximated by means of an analytic function, while in the second step the analytical derivative of the approximating function is computed.

Let us first consider the 1D case. If g_j is the datum at the location x^j, $j = 1, \ldots, N$, the approximating function is the function f that minimizes the functional

$$\Psi_\lambda[f] = \sum_{j=1}^{N} (g_j - f(x^j))^2 + \lambda \int (f''(x))^2 \, dx. \qquad (5.18)$$

with $\lambda > 0$. Clearly, the regularizing functional

$$\| Cf \|^2 = \int (f''(x))^2 \, dx$$

ensures the smoothness of the approximating function.

Under rather general assumptions (Poggio *et al.*, 1984), it can be shown that the function f that minimizes the functional Ψ_λ of eqn (5.18) can be obtained by convolving the data with an appropriate filter R. Therefore, the derivative of f can be computed by convolving the data with the derivative of the filter.

In the 2D case, if the regularizing functional is

$$\| Cf \| = \iint (\nabla^2 \operatorname{grad} f)^2 \, dx_1 \, dx_2$$

where ∇^2 is the Laplacian, it can be shown (Poggio *et al.*, 1984) that the solution can be obtained by convolving the data with the filter

$$R_2(x_1, x_2) = \frac{1}{2} \int_0^\infty \frac{J_0(\omega z)}{\lambda \omega^6 + 1} \, \omega \, d\omega$$

where J_0 is the 0th-order Bessel function and $z = \sqrt{x_1^2 + x_2^2}$.

In practice, the convolution of the data with a low-pass filter is sufficient to regularize the problem of differentiation. Consequently, most of the proposed methods for edge detection, methods developed before the connection between early vision and regularization theory was pointed out, can be regarded as regularization methods.

Let us now deal with the problem of the computation of optical flow.

5.4.2 Computing optical flow

In Section 5.2 it was shown that the problem of determining the optical flow from eqn (5.9) is ill-posed. The well-posedness can be restored by means of the notion of generalized solution. Let us rewrite eqn (5.9) as

$$\operatorname{grad} E \cdot v = -E_t. \tag{5.19}$$

The pseudosolutions to eqn (5.19) are the vector fields which minimize the functional

$$\| \operatorname{grad} E \cdot v + E_t \|.$$

If $\operatorname{grad} E \neq 0$, a generic vector field v can be uniquely decomposed into the pair $(v_\perp, v_\parallel)$, with $v_\perp$ the component of v in the direction orthogonal to the isobrightness contour (that is, parallel to $\operatorname{grad} E$), and $v_\parallel$ the component of v in the direction parallel to the isobrightness contour.

From the decomposition $v = (v_\perp, v_\parallel)$, it follows that all the pseudosolutions to eqn (5.19) are the vector fields v with

$$v_\perp = -\frac{E_t}{\| \operatorname{grad} E \|}.$$

The normal pseudosolution is the unique pseudosolution v^+ with $v_\parallel = 0$. Interestingly, the vector field v^+ can be successfully employed for motion understanding (Aloimonos and Duric, 1994).

In the search for a unique optical flow more similar to the perceived image motion, Horn and Schunck (1981) proposed to look for the vector field $v = (v_1, v_2)$ which minimizes the functional

$$\Omega_\lambda[v] = \int\!\!\int [(dE/dt)^2 + \lambda(|\operatorname{grad} v_1|^2 + |\operatorname{grad} v_2|^2)] \, dx_1 \, dx_2 \tag{5.20}$$

with $\lambda > 0$ and where the integral extends over the whole image plane. The second term in the right-hand side of eqn (5.20) is a smoothness constraint and, hence, the problem of minimizing $\Omega_\lambda[v]$ is well posed. Intuitively, the unique solution v is the smoothest vector field that nearly satisfies eqn (5.8).

The iterative scheme proposed by Horn and Schunck for the minimization of $\Omega_\lambda[v]$ makes it clear that the solution propagates from the boundary to the interior of the image plane.

An interesting generalization of the smoothness term of $\Omega_\lambda[v]$ can be found in (Youille and Grzywacz, 1988) where the optical flow is computed as the vector field v which minimizes the functional

$$\hat{\Omega}_\lambda[v] = \int\!\!\int \left[(dE/dt)^2 + \lambda \sum_{r=0}^{\infty} c_r (D^r v)^2 \right] dx_1 \, dx_2 \tag{5.21}$$

with $\lambda > 0$,

$$c_r = \frac{\sigma^{2r}}{r! 2^r},$$

and

$$(D^{2r}v) = \nabla^{2r}v$$

$$(D^{2r+1}v) = \text{grad}\,(D^{2r}v)$$

(∇^2 is the Laplacian operator). The functional $\hat{\Omega}_\lambda$ of eqn (5.21) has three interesting properties (Youille and Grzywacz, 1988). First, the fact that $c_0 > 0$ is a necessary and sufficient condition for the interaction to fall faster than $1/r$, where r is the distance between motion measurement sites. Second, due to the particular choice of the coefficients c_r, the smoothing effects of the regularizing term in $\hat{\Omega}_\lambda$ is equivalent to a Gaussian interaction. Surprisingly enough, this 'short range' interaction, which is more effective than the 'long range' inter-action of the Horn and Schunck method, is induced by the presence of higher-order derivatives in the regularizing functional. Third, the optical flow which is obtained by minimizing $\hat{\Omega}_\lambda$ seems to be consistent with a number of psychophysical experiments on motion perception.

Finally, let us turn to the regularization of shape from shading.

5.4.3   Shape from shading

In Section 5.2 it was shown that the problem of solving eqn (5.12) is ill posed. In order to recover shape from shading (Ikeuchi and Horn, 1981) proposed to minimize the functional

$$\Theta_\lambda[p, q] = \int\int [|E(x) - R(p, q)|^2 + \lambda(|\text{grad}\,p|^2 + |\text{grad}\,q|^2)]\,dx_1\,dx_2 \quad (5.22)$$

with $\lambda > 0$ and where

$$R(p, q) = s \cdot n$$

is the reflectance map of a Lambertian surface, with s the unit vector in the direction of the light source, and

$$n = \frac{1}{\sqrt{1 + p^2 + q^2}}\,(-p, -q, 1)$$

the unit vector normal to the surface $f = f(x_1, x_2)$.

Since the reflectance map depends nonlinearly on p and q, the problem of minimizing the functional Θ_λ of eqn (5.22) is not necessarily well posed. A rigorous proof of the existence and continuous dependence of the solution on the data of this problem can be found in (Bertero *et al.*, 1988). The minimization of Ω_λ, usually obtained by means of iterative schemes, poses an interesting question. The devised iterative algorithms tend to 'walk away' from the correct solution of the image irradiance equation when this solution is provided as the initial condition. In essence this somewhat surprising behavior is the typical side effect of regularization techniques: a small amount of error in the original equation is traded for an increase in the smoothness of the solution.

From the solutions $p = p(x_1, x_2)$ and $q = q(x_1, x_2)$, the viewed surface $f = f(x_1, x_2)$ can then be recovered as the function that minimizes the functional

$$\int\int [(f_{x_1} - p)^2 + (f_{x_2} - q)^2] \, dx_1 \, dx_2$$

with the appropriate boundary conditions (Ikeuchi and Horn, 1981).

In order to obtain more faithful surface reconstruction, Horn (1990) suggested the minimization of the functional

$$\hat{\Theta}_{\lambda,\mu}[p, q, f] = \int\int \{|E(x) - R(p, q)|^2 + \lambda(|\operatorname{grad} p|^2 + |\operatorname{grad} q|^2)$$
$$+ \mu[(f_{x_1} - p)^2 + (f_{x_2} - q)^2]\} \, dx_1 \, dx_2 \qquad (5.23)$$

with $\lambda, \mu > 0$. The explicit presence of a term which measures the error of surface reconstruction in eqn (5.23), reduces the 'walk away' effect from the correct solution of iterative techniques and appears to produce better results on real images (Horn, 1990).

5.5 DISCUSSION

In this section the impact of regularization theory in computer vision is discussed. Let us start by listing the merits of the described framework.

5.5.1 A unified approach to early vision

First, the regularization theory approach to computer vision provides a coherent framework in which many visual problems can be analysed. As shown in Sections 5.2 and 5.4 several early vision problems can be formulated and solved as particular instances of a general class of problems, the class of ill-posed problems.

Second, the connection between vision and regularization theory clarified important aspects of the variational formulation of visual problems (like the role of smoothness constraints and *a priori* information in the search for satisfactory solutions (Youille and Grzywacz, 1988), and stimulated the development of many fruitful ideas (like controlled continuity constraints

(Terzopoulos, 1986), snakes (Kass *et al.*, 1987), balloons (Cohen and Cohen, 1993), and deformable contours (Blake *et al.*, 1993)).

Third, a number of interesting works originated from the attempt to overcome the difficulty of regularization to dealing with discontinuity. A largely incomplete list includes papers on Markov random fields and stochastic relaxation techniques for image segmentation (Geman and Geman, 1984; Poggio *et al.*, 1988; Geman *et al.*, 1990) and the solution to variational problems with discontinuities (Blake and Zisserman, 1987; Lee and Pavlidis, 1988; Mumford and Shah, 1989; Ambrosio and Tortorelli, 1990; March, 1992). Let us now turn to the major criticism that can be raised against the described approach.

5.5.2 Is early vision really ill posed?

The need of regularization theory for computer vision was advocated on the basis of the inverse nature of many early vision problems. This point bears reflection because the ill-posedness of an inverse problem might depend on the specific choice of the X and Y spaces and on the available *a priori* information on the solution.

The *uniqueness* condition, for example, depends critically on the adopted (and available) geometric and heuristic constraints that are used to resolve ambiguities. The computation of image motion provides an instructive example. If the image brightness constancy equation is assumed to be the only available constraint, then the computation of optical flow is undoubtedly ambiguous. However, it has been shown that other constraint equations or suitable assumptions on the local structure of the optical flow of rigid objects (Tretiak and Pastor, 1984; Uras *et al.*, 1988; Verri *et al.*, 1990) can be used to efficiently and successfully reconstruct the full 2D motion field. In many cases, these constraints might be better suited than regularizing functionals for the description of the structural properties of the 2D motion field.

In short the extent to which the full machinery of regularization theory is really necessary and useful for computer vision seems to depend on the amount of *a priori* information actually available for each specific problem.

Finally, let us summarize the content of this chapter.

5.6 CONCLUSIONS

Regularization theory studies methods for the solution to ill-posed problems (i.e., problems for which at least one of the conditions of *uniqueness, existence,* or *continuous dependence* of the solution *on the data* is not ensured). Many early vision problems can be formulated as problems of inverse optics and appear to be ill posed.

In this chapter a few methods which have been proposed for the solution to classical early vision problems, like edge detection, motion estimation, and shape from shading have been discussed in the light of regularization theory. It is concluded that regularization theory provided a unified and inspiring framework for computational studies of early vision.

REFERENCES

Adelson, E.H. and Bergen, J.R. (1985) Spatiotemporal energy models for the perception of motion. *J. Opt. Soc. Am.* **A2**, 284–299.

Aloimonos, Y. and Duric, Z. (1994) Estimating the heading direction using normal flow. *Int. Journal of Computer Vision* **13**, 33–56.

Ambrosio, L. and Tortorelli, V.M. (1990) Approximation of functionals depending on jumps by elliptic functionals via Γ-convergence. *Commun. Pure & Applied Mathematics* **43**, 999–1036.

Bertero, M. (1989) linear inverse and ill-posed problems. *Advances in Electronics and Electron Physics* **75**, 1–120.

Bertero, M., Poggio, T. and Torre, V. (1988) Ill-posed problems in early vision. *Proc. IEEE* **76**, 869–889.

Blake, A. and Zisserman, A. (1987) *Visual reconstruction*. MIT Press, Cambridge, Massachusetts.

Blake, A., Curwen, R. and Zisserman, A. (1993) A framework for spatiotemporal control in the tracking of visual contours. *Int. J. Computer Vision* **11**, 127–145.

Born, M. and Wolf, E. (1959) *Principles of optics*. Pergamon Press, New York.

Canny, J.F. (1983) Finding edges and lines in images. *AI Lab Memo* **720**, MIT, Cambridge, Massachusetts.

Cohen, L.D. and Cohen, I. (1993) Finite-element methods for active contour models and ballons for 2D and 3D images. *IEEE Trans. Pattern Analysis Machine Intelligence* **PAMI-15**, 1131–1147.

Courant, R. and Hilbert, D. (1962) *Methods of mathematical physics (II)*. Wiley Interscience, London, UK.

Craven, P. and Wahba, G. (1979) Smoothing noisy data with spline functions: estimating the correct degree of smoothing by the method of generalized cross validation. *Numerical Mathematics* **31**, 377–403.

Geman, S. and Geman, D. (1984) Stochastic relaxation, Gibbs distributions and the Bayesian restoration of images. *IEEE Trans. Pattern Analysis Machine Intelligence* **PAMI-6**, 721–741.

Geman, D., Geman, S., Graffigne, C. and Dong, P. (1990) Boundary detection by constrained optimization. *IEEE Trans. Pattern Analysis Machine Intelligence* **PAMI-12**, 609–628.

Hadamard, J. (1902) Sur les problèmes aux dérivées partielles et leur signification physique. *Princeton University Bulletin* Vol. **13**.

Heeger, D.J. (1987) Optical flow using spatiotemporal filters. *Int. J. Computer Vision* **1**, 279–302.

Heeger, D.J. and Jepson, A.D. (1992) Subspace methods for recovering rigid motion I: algorithm and implementation. *Int. J. Computer Vision* **7**, 95–117.

Hildreth, E.C. (1984) The computation of the velocity field. *Proc. Royal Society of London* **B221**, 189–220.

Horn, B.K.P. (1975) Obtaining shape from shading information. In: *The psychology of computer vision* (ed. P.H. Winston). McGraw-Hill, New York.

Horn, B.K.P. (1990) Height and gradient from shading. *Int. Journal of Computer Vision* **5**, 37–75.

Horn, B.K.P. and Schunck, B.G. (1981) Determining optical flow. *Artificial Intelligence* **17**, 185–203.

Horn, B.K.P. and Sjoberg, R.W. (1979) Calculating the reflectance map. *Applied Optics* **18**, 1770–1779.

Ikeuchi, K. and Horn, B.K.P. (1981) Numerical shape from shading and occluding boundaries. *Artificial Intelligence* **17**, 141–184.

Kass, M., Witkin, A. and Terzopoulos, D. (1987) Snakes: active contour models. *Int. Journal of Computer Vision* **1**, 321–331.

Lee, D. and Pavlidis, T. (1988) One dimensional regularization with discontinuities. *IEEE Trans. Pattern Analysis and Machine Intelligence* **PAMI-10**, 822–829.

Longuet-Higgins, H.C. and Prazdny, K. (1981) The interpretation of moving retinal images. *Proc. Royal Society of London* **B208**, 385–397.

March, R. (1992) Visual reconstruction with discontinuities using variational methods. *Image and Vision Computing* **10**, 30–38.

Mumford, D. and Shah, J. (1989) Optimal approximations by piecewise smooth functions and associated variational problems. *Commun. Pure & Applied Mathematics* **42**, 577–685.

Mundy, J.L. and Zisserman, A. (1992) Appendix – Projective geometry for machine vision. In: *Geometric invariants in computer vision* (eds J.L. Mundy and A. Zisserman), MIT Press, Cambridge, Massachusetts.

Poggio, T., Voorhees, H. and Yuille, A. (1984) Regularizing edge detection. *AI Lab. Memo* **776**, MIT, Cambridge, Massachusetts.

Poggio, T., Torre, V. and Koch, C. (1985) Computational vision and regularization theory. *Nature* **317**, 314–319.

Poggio, T., Gamble, E.B. and Little, J.J. (1988) Parallel integration of vision modules. *Science* **242**, 436–440.

Rieger, J.H. and Lawton, D.T. (1985) Processing differential image motion. *J. Optical Society of America* **A2**, 354–359.

Shanmugam, K.F., Dickey, F.M. and Green, J.A. (1965) An optimal frequency domain filter for edge detection in digital pictures. *IEEE Trans. Pattern Analysis Machine Intelligence* **44**, 99–149.

Terzopoulos, D. (1986) Regularization of inverse visual problems involving discontinuities. *IEEE Trans. Pattern Analysis Machine Intelligence* **PAMI-8**, 413–424.

Tikhonov, A.N. and Arsenin, V.Y. (1977) *Solutions of ill-posed problems.* Winston & Sons, Washington, DC, USA.

Torre, V. and Poggio, T. (1986) On edge detection. *IEEE Trans. Pattern Analysis Machine Intelligence* **PAMI-8**, 147–163.

Tretiak, O. and Pastor, L. (1984) Velocity estimation from image sequences with second order differential operators. *Proc. Int. Conf. on Pattern Recognition*, Montreal, Canada, 16–19.

Uras, S., Girosi, F., Verri, A. and Torre, V. (1988) A computational approach to motion perception. *Biological Cybernetics* **60**, 79–87.

Verri, A. and Poggio, T. (1989) Motion field and optical flow: qualitative properties. *IEEE Trans. Pattern Analysis and Machine Intelligence* **PAMI-11,** 490–498.

Verri, A., Girosi, F. and Torre, V. (1990) Differential techniques for optical flow. *J. Optical Society of America* **A7,** 912–922.

Wahba, G. (1977) Practical approximate solutions to linear operator equations when the data are noisy. *SIAM J. Numerical Analysis* **14**.

Youille, A. and Grzywacz, N.M. (1988) A computational theory for the perception of coherent visual motion. *Nature* **333,** 71–73.

PART II

INTEGRATING VISUAL MODULES

An increasing number of real-world applications involve visual modules as components of more complex intelligent systems. The latter maintain hypotheses and take decisions about the acquired data, on the basis of internal models, possibly synthesized (learned) from the same external data stream. Problems to the designers of such model-based systems arise from the need of processing large amounts of noisy, ambiguous and time-varying data.

Under this perspective, a number of topics become of primary concern: building (abstracting) models, by combining suited primitive elements (e.g., numerical, geometric, linguistic); indexing models, and matching them against the acquired data; fusing sensory data into more robust descriptions; propagating the uncertainty, in a consistent way, up to final hypotheses; integrating modules, by means of distributed architectures and communication supports.

Such topics are addressed in the present part of the volume.

Chapter 6 discusses geometrical modelling techniques for synthesizing 3D object descriptions. Object models are seen as basic structures to enable spatial reasoning for image understanding and robotics. Chapter 7 introduces a specific distributed scheme – the probabilistic or Bayes networks – which provides a unified framework for both the fusion and the treatment of uncertain patterns. Propagating uncertainty in such a scheme is actually a consistent technique of automated reasoning. Chapter 8 gives an overview on the main architectural solutions proposed to support the fusion of information (data, decisions) in visual systems. The support is provided by distributed systems, which are extensively discussed. Finally, Chapter 9 addresses the problem of fusing subsymbolic with symbolic descriptions, and presents a hybrid architecture that integrates learning – through a connectionist module – with reasoning about the scene at a symbolic/linguistic level.

6

Geometric Modelling and Spatial Reasoning

6.1   INTRODUCTION

We consider the problem of modelling and manipulating 3D shapes, with a special emphasis on applications in image understanding. The problem of designing a representation scheme for solid objects has been extensively studied in the context of CAD applications. The object representation problem played an important role in image understanding as well.

Most computer vision systems, which are designed to recognize three-dimensional objects, compare a *scene model*, constructed by processing images obtained from one or more sensors, against entities in a *model database* containing a description of each object the system is expected to recognize (Flynn and Jain, 1991). The development of such *model-based* recognition techniques has occupied the attention of many researchers in the image understanding community for years (Besl and Jain, 1985; Chin and Dyer, 1986; Brady *et al.*, 1988). Three-dimensional model-based image understanding

ARTIFICIAL VISION
ISBN 0-12-444816-X

Copyright © 1997 Academic Press Ltd
All rights of reproduction in any form reserved

uses geometric models and sensor data to recognize objects in a scene. Likewise, CAD systems are used to interactively generate 3D object models during the design process. In recent years, the unification of CAD and vision systems has become the focus of research in the context of manufacturing automation (Hansen and Henderson, 1989).

Computer vision researchers have used a variety of model types, which can be broadly classified as *quantitative* (i.e., the model can be used to generate a synthetic image of the object), or *qualitative* (i.e., the model information can be used to distinguish between different objects, but not to generate synthetic images). Solid representations used in CAD are quantitative in nature: primitives (surfaces and volumes) are specified in terms of numerical parameters. Instead, if we are performing a visual recognition task and the objects to be recognized can be distinguished by examining some qualitative features of the segmented primitives, representations, which capture only those variations, might be advantageous (Flynn and Jain, 1991).

In the paper, after analysing representation schemes for solid objects to be used during a recognition process, we consider geometric problems which arise when spatial reasoning techniques are applied in an image understanding context. In particular, we show through examples how geometric algorithms developed in the field of computational geometry have been successfully employed for solving image understanding problems in application domains like robotics and biomedicine. Such problems include the reconstruction of a 3D shape from points or segments on its boundary. Furthermore, we show the wide applicability of geometric structures like the Delaunay triangulation in two or three dimensions and of the corresponding construction algorithms in such contexts. The remainder of the paper is organized as follows. Section 6.2 deals with issues arising in designing object representations suitable for recognition. In Section 6.3, a classification and description of object representations is provided, which basically distinguishes between boundary and volumetric schemes. In Section 6.4, we discuss modelling issues in image understanding, with special attention to multiview descriptions. Finally, Section 6.5 is devoted to examples of geometric structures and algorithms used for solving problems typical of spatial reasoning.

6.2 OBJECT REPRESENTATIONS FOR RECOGNITION

In this section, we consider representation issues which are important during the recognition process. Many different approaches have been proposed to represent a three-dimensional object in a form suitable for recognition (Binford, 1984; Besl and Jain, 1985; Besl, 1988; Chin and Dyer, 1986), but a general-purpose modelling scheme, i.e., a systematic approach for building models for a large class of objects has not emerged yet. A lot of work has been performed on

2D (image-space) and $2\frac{1}{2}$D (surface-space) representations of three-dimensional scenes (Chin and Dyer, 1986). Both 2D and $2\frac{1}{2}$D representations provide *viewer-centered* object descriptions. $2\frac{1}{2}$D representations use descriptions of surfaces instead of boundaries, and therefore have the advantage of more accurately representing the complete object, thus improving chances for reliable recognition. However, if we assume that an object can occur in a scene at an arbitrary orientation in the three-dimensional space, then the model must contain a description of the object from *all* viewing angles. In addition, in an industrial automation environment, in which the vision system must be integrated with a CAD database, a three-dimensional model may be convenient because of its compatibility.

The design of a representation scheme for object recognition is dominated by three issues: (i) the method adopted for acquiring the object representation; (ii) the choice of a suitable representation; (iii) the choice of a reference frame for the representation.

Methods for acquiring object representations have received relatively little attention. Usually, representations used for recognition have been provided manually. Manual construction of object descriptions is obviously time consuming and requires detailed knowledge of the object recognition system. Thus, this approach is impractical in applications where the set of objects to be recognized is large or changes frequently. An alternative is to construct representations from examples by using prototypical features extracted from images taken from a number of viewpoints: representations built by learning are limited in their precision by the quality of the sensor. Much interest centers on using CAD tools to construct object representations for recognition. Some authors, for instance, use commercial CAD systems for generating the object models to be stored in the database and augment such description with geometric knowledge necessary for recognition (Flynn and Jain, 1991; Hansen and Henderson, 1989; Zhang *et al.*, 1993).

The attributes and the geometric relations among component parts of an object must be defined with respect to a coordinate system, which in vision research is either *object-* or *viewer-centered*. Most work in object recognition has adopted object-centered descriptions, since they provide a natural way to express objects independently of the point of view (Brooks, 1981; Marr and Nishihara, 1978). However, the non-linear transformation between the viewer-centered coordinate frame and the object-centered one must be determined. Object-centered representations have therefore been best suited to problems where the location and orientation of the object relative to the viewer is known approximately, or can be discovered by some relatively simple means (Zhang *et al.*, 1993). Recently, there has been a lot of work on multiview descriptions, which describe objects through a finite set of viewer-centered descriptions (Korn and Dyer, 1987): each member of the set represents the object as its 2D projection as seen from one viewpoint. By using such representations, features extracted from images can be directly matched with features associated with each member of the multiview model set. The shift from object-centered descriptions to viewer-centered ones shifts processing requirements from object recognition time to object modelling time. However, as the number of views in the set of 3D multiview descriptions increases, the storage space and matching time required increase as well.

In the literature, the word 'model' is overloaded, in the sense that it is used both for indicating an abstraction of a physical object and for a representation of it. It is possible to formally define the problem of modelling physical objects by introducing three sets of entities, called *spaces* (Requicha, 1980): the *physical object space* S_{ph}, the *mathematical modelling space* S_{math} and the *representation space* S_{rep}. S_{ph} defines the three-dimensional real world we want to model. To model a real-world entity within a computer, we must adopt a suitable idealization of such entity; thus, we define a class of mathematical objects, characterized by using point set theory and algebraic topology (Mäntylä, 1987), which define S_{math}. Then, we want to assign to a mathematical object a suitable representation within a computer, i.e., to describe it as a finite collection of symbols (of a finite alphabet): the set of the representations define S_{rep}.

The most general mathematical abstraction of a real solid object is a subset of the three-dimensional Euclidean space $\mathbb{E}^3$. Very few subsets of $\mathbb{E}^3$ are adequate models of solid objects. A mathematical model of a solid object should capture properties which correspond to our intuitive notion of a solid (Requicha, 1980), such as finiteness, rigidity, homogeneous three-dimensionality, boundary determinism and finite describability. Mathematical models satisfying such properties are *r-sets* which are subsets A of $\mathbb{E}^3$ such that $A \equiv \mathring{\bar{A}}$[1], i.e., subsets of $\mathbb{E}^3$ without any lower-dimensional dangling part. The modelling space is sometimes restricted to *three-manifolds*; a three-manifold is a subset of $\mathbb{E}^3$ where each point has a neighborhood homeomorphic to a sphere. In what follows, we assume that mathematical models are those defined by three-manifold solids, since representation schemes developed in the literature are based on such assumption. Once the mathematical modelling space is fixed, each physical object can be identified with its model. Hence, we will use the terms solid object and model interchangeably.

A *representation scheme* is formally defined as a relation $\gamma : S_{math} \rightarrow S_{rep}$. The *domain* of γ is denoted by D_γ and its image under γ by V_γ. Any representation in V_γ is called *valid*. However, we neither assume that all models are representable (i.e., in general $S_{math} \neq D_\gamma$) nor that all representations of S_{rep} are valid. A valid representation r is *unambiguous* (or *complete*) if it corresponds to a single object, it is *unique* if its corresponding object in S_{math} does not admit any representation different from r.

---

[1] Here, $\mathring{A}$ and $\bar{A}$ denote the interior and the closure of a set A, respectively.

6.3 REPRESENTATION SCHEMES

A general classification of representation schemes, originated in the CAD literature, distinguishes between *boundary* and *volumetric* schemes. A boundary representation scheme describes a solid object in terms of the surfaces enclosing it, while a volumetric representation scheme describes an object in terms of solid primitives covering its volume. Volumetric representations can be classified into:

- *decomposition* representations, which describe an object in terms of a collection of primitive objects combined through a gluing operation;
- *constructive* representations, which describe an object as the Boolean combination of primitive point sets;
- *geon-based* representations, which describe an object as a set of primitives plus a set of spatial connectivity relations among them.

6.3.1 Boundary schemes

A *boundary representation* (BRep) of an object is a geometric and topological description of its boundary. The object boundary is segmented into a finite number of bounded subsets, called *faces*. A face is represented in a BRep by its bounding *edges* and *vertices*. Thus, a BRep consists of three primitive topological entities: *faces* (2-dimensional entities), *edges* (1-dimensional entities) and *vertices* (0-dimensional entities).

Geometric information consist of the shape and location in space of each of the primitive topological entities.

There are nine different topological relations involving faces, edges and vertices: each relation consists of a unique ordered pair from the three topological entities (faces, edges and vertices) (Weiler, 1985). Such relations are subdivided into:

- *adjacency* relations, which relate an entity of dimension k ($k = 1, 2$, i.e., edges and faces) with all k-dimensional entities sharing a $(k-1)$-dimensional entity with it; for $k = 0$, we say that two vertices are adjacent if they are endpoints of the same edge;
- *incidence* relations, which relate an entity of dimension k ($0 \leqslant k \leqslant 2$) with all m-dimensional ($0 \leqslant m \leqslant 2$) entities incident at it, with $m \neq k$; such relations are further classified into:
  - *incidence coboundary* relations, if $m > k$;
  - *incidence boundary* relations, if $m < k$.

Note that, for $k = 2$, only adjacency and incidence boundary relations are defined, while, if $k = 0$, only adjacency and incidence coboundary relations exist.

In a data structure for describing a BRep, the three topological entities must be stored together with a subset of the nine topological relations. The idea is to store a subset of the nine relations which allows a complete reconstruction of the non-stored relations without errors or ambiguities (Weiler, 1986). The resulting data structure is called *sufficient*. A data structure is considered *efficient* if each non-stored relation can be retrieved from the data structure in a time proportional to the number of entities involved in the relation itself. Several data structures have been proposed in the literature to encode a BRep, which differ in the entities and relations they store: the *winged-edge structure* (Baumgart, 1972), the *symmetric structure* (Woo, 1985), the *face adjacency hypergraph* (Ansaldi *et al.*, 1985), the *quad-edge structure* (Guibas and Stolfi, 1985), the *vertex-edge* and *face-edge* data structures (Weiler, 1986), the *half-edge* structure (Mäntylä, 1987).

Boundary representation schemes depend on the surfaces that can be used. Thus, boundary representation schemes can represent a wide variety of solid objects at arbitrary levels of detail; they are unambiguous, if faces are explicitly represented, but generally they are not unique. Boundary representations are usually quite verbose, especially when curved-faced objects are approximated by polyhedral representations, called *faceted representations*. Boundary schemes are especially useful to generate graphical outputs, because of the availability of boundary information, and for describing tolerances (i.e., information attached to the boundary entities of an object). Integral properties can be easily and efficiently computed from a BRep when operating in a planar-faced environment. Boolean set operations are costly and tedious to implement on BReps, although they are simplified for faceted representations, especially when facets have a simple geometry (like triangles). As we will point out in Section 6.4, BReps are suitable for image understanding applications, since they provide a description of the boundary of an object, which is related to the appearance of an object in an image.

If the domain of the mathematical models is extended from three-manifolds to *r*-sets, the data structures previously described can still be used, although some modifications must be applied.

If the domain is further extended to non-manifolds subsets of $\mathbb{E}^3$ with dangling faces or edges, it is necessary to use *non-manifold boundary schemes*, which have been originated as an extension of BReps (Weiler, 1986; Gursoz *et al.*, 1990; Murabata and Higashi, 1990).

6.3.2 Decomposition schemes

Decomposition representations are volumetric models which describe an object as a collection of simple primitive objects combined with a single 'gluing' operation. Decomposition schemes can be classified into *object-based* and *space-based* schemes. The former describe an object as the combination of pairwise quasi-disjoint elementary 3D cells whose union covers the object, and are known as *cell complexes* (or *cellular decompositions*). The latter decompose the 3D space occupied by the object into elementary volumes (usually cubes), and represent the object as the combination of the volume elements belonging to it: they include spatial enumeration and space-based adaptive subdivision schemes.

According to their general definition *cell complexes* are collections of k-dimensional cells $(k = 0, 1, 2, 3)$ whose union is an r-set. A k-dimensional cell usually is any subset of $\mathbb{E}^n$ $(n > k)$ topologically equivalent to a k-dimensional sphere. The boundary of each k-dimensional cell $(k > 0)$ is the union of cells which have a dimension lower than k. Each k-dimensional cell $(k < 3)$ belongs to the boundary of a $(k + 1)$-dimensional cell. In a cell complex, pairs of three-dimensional cells must be either completely disjoint, or meet at one vertex (0-cell), or along an edge (1-cell), or along a face (2-cell). When the cells are simplices, i.e. each k-cell is a linear convex combination of $k + 1$ points, the cell complex is said to be a *simplicial complex*. In three dimensions a simplex is a tetrahedron. Then, a simplicial complex in the space is usually called *tetrahedralization* (see Figure 6.1).

Several topological data structures have been defined in recent years describing subdivisions, 3-dimensional and even n-dimensional cell complexes. According to Lienhardt (1991), there are basically two types of data structures for encoding cell complexes:

1) *incidence graphs*, which are graphs or hypergraphs, where the nodes correspond to the cells of the subdivision and the arcs to the adjacency/ incidence relations among the cells (Edelsbrunner, 1987; Rossignac, 1991);
2) *oriented topological structures*, which use a single type of basic elements and a set of element functions which act on their elements (Dobkin and Laszlo, 1987; Brisson, 1989).

Data structures for describing the boundary of solid objects (i.e., 2-dimensional cell complexes) have been described in Section 6.3.1. A survey of

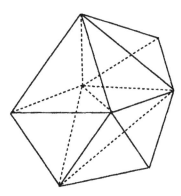

Figure 6.1 An example of 3D simplicial complex (tetrahedralization).

data structures for describing cell complexes in *n*-dimensions can be found in Brisson (1989). A special data structure for encoding simplicial complexes is described in Ferrucci and Paoluzzi (1991).

The expressive power of cell complexes is quite large, since the modelling space is general. They are unambiguous representation schemes, but not unique. Cell complexes have several important properties, like invariance through rigid transformations, ease of update and computational efficiency. Usually, these representations are created by conversion from another solid model or by automatic procedures which start from sparse sets of points or segments in the 3D space. Such schemes are mainly used for object reconstruction (Boissonnat, 1982), and finite element meshes.

Space-based decomposition schemes describe objects in terms of the space they occupy by subdividing the 3D space into regular volume elements, called *voxels*. An object is represented as the collection of the voxels which have a non-empty intersection with the object itself. The first space-based scheme, which is a generalization of the pixel-based schemes used in image processing, is the *spatial enumeration* scheme. In such a scheme, a representation of an object *S* is a list of spatial cells occupied by *S*. The cells (voxels) are cubes of fixed size and lie in a fixed spatial grid, thus forming a regular subdivision of the space. Spatial enumeration representation schemes are unambiguous. They are not invariant under translations and rotations and tend to be quite verbose: also, storage costs dramatically increase with increased resolution. A great advantage of such representations is the conceptual simplicity of Boolean operations on them. The main disadvantages is in the fact that they produce only approximate object descriptions, where the quality of the approximation is determined by the voxel size.

To reduce storage requirements, adaptive subdivision schemes have been developed like the *octree* or the *bintree* (Jackins and Tanimoto, 1984; Samet, 1990a; Samet, 1990b; Tamminen, 1984). An octree (Jackins and Tanimoto, 1984; Samet, 1990a; Samet, 1990b) is the three-dimensional generalization of the quadtree used for representing two-dimensional entities. The octree uses a recursive regular subdivision of a cubic universe into eight octants. An octant is not subdivided when it is completely inside or outside the object, or its size is equal to the resolution (voxel level). This recursive subdivision is described by a tree of degree eight, where the root represents the universe, and each non-root node is a cube obtained from the subdivision of its parent. Leaves describe portions of space for which no further subdivision is necessary or possible (voxels).

Much work has been done in developing efficient data structures and algorithms for storing and processing quadtrees and octrees. The two books by Samet are excellent references for these subjects (Samet, 1990a; Samet, 1990b). As spatial enumeration schemes, octrees are approximate representations: the quality of the approximation is determined by a fixed resolution. Efficient algorithms, based on a single tree traversal, have been developed to perform Boolean operations on octrees and to compute volume and integral properties efficiently. Alternative octree-based representations (for instance, the PM-octrees) have been proposed, which overcome the limit of providing only an approximate (resolution-dependent) description, which is typical of octrees.

6.3.3 Constructive schemes

Constructive Solid Geometry (CSG) (Requicha, 1980; Requicha and Voelcker, 1982; Requicha and Voelcker, 1983) defines a family of schemes for representing solids as Boolean combination of predefined volumetric primitives. The most natural way to represent a CSG model is the so-called *CSG tree*. A CSG tree is a binary tree in which internal nodes represent operators, which can be either rigid motions or regularized union, intersection or difference, while terminal nodes are either primitive leaves which represent subsets of $\mathbb{E}^3$, or transformation leaves which contain the defining arguments of rigid motions. Each subtree, that is not a transformation leaf, represents a set resulting from applying the indicated operators to the sets represented by primitive leaves. There are two kinds of CSG schemes depending on whether the primitives are *r*-sets (called *CSG*), or are half-spaces (*CSG-based on general half-spaces*). The former is the most common, since for human users it is easier to operate with bounded primitives than with unbounded half-spaces. Figure 6.2 shows an example of a CSG tree (with bounded primitives).

A CSG representation is a *procedural* (or *unevaluated*) scheme, since it does not describe explicitly, unlike boundary or decomposition schemes. CSG representation schemes are unambiguous, but clearly not unique. The expressive power of a CSG scheme depends on the set of primitive solids and on the motional and Boolean operators available. If the primitives of a CSG scheme are bounded, then any CSG tree is a valid representation if the primitive leaves are valid.

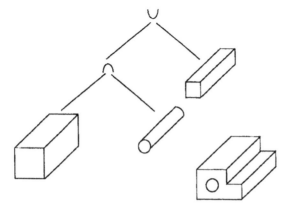

Figure 6.2 A CSG representation.

CSG schemes are typical of a design environment, since they somehow simulate the process of object construction by a designer. The main disadvantage of a CSG representation is the difficulty in computing integral properties and in extracting or describing information related to surfaces. Moreover, CAD systems based on a CSG representation always store a boundary description of the object to speed up rendering as well as analysis operations.

6.3.4 Geon-based schemes

Unlike decomposition and constructive schemes, geon-based schemes are qualitative volumetric representations.

A qualitative representation was first proposed by Biederman (1985) as a model of human vision, but it also offers interesting properties for a computer model. In human vision, the retinal image is transformed at different levels of the visual pathway into various data representations as a precursor to a possible object recognition. At the highest levels in this process, we have only a very sketchy knowledge of the exact details (Levine *et al.*, 1991). The main idea behind Biederman's approach is to coarsely reconstruct 3D objects using generic primitives, called *geons*. Biederman developed a catalog of 36 *geons* with each member of the catalog having a unique set of four qualitative features (Flynn and Jain, 1991):

- *edge*: straight or curved;
- *symmetry*: rotational/reflective, reflective, asymmetric;
- *size variations*: constant, expanding, expanding/contracting;
- *axis*: straight or curved.

Geons can be seen as a symbolic characterization of a subset of simple shapes from a generalized cylinders representation. They characterize the individual parts of an object. In a geon-based scheme, an object is then described by combining the geon shapes and spatial relations among these geons. Figure 6.3 shows an example of geon representation for primitive volumetric parts. The object is thus modelled as an *object graph* $G = (N, A)$ (Levine *et al.*, 1991), where N is a set of nodes representing parts, each node being assigned a unique geon type, and A is a set of directed arcs representing their spatial relations (which are essentially connection relations). The object graph is obtained from a description in the form of a graph of the original image (for instance, a line-junction graph).

The expressive power of geon-based representations is quite high. Over 154 million qualitatively different objects can be constructed from three geons and intergeon relationships, considering all possible relative sizes, arrangements and affixments of the three components (Biederman, 1987). Biederman's geons constitute only one possible selection of qualitatively-defined volumetric primitives; the general approach of applying the Cartesian product to a set of

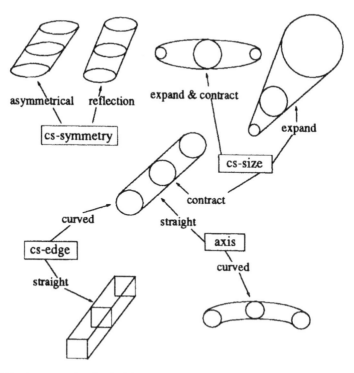

Figure 6.3   An example of primitive volumetric parts in a geon-based representation.

contrastive primitive properties (e.g., straight versus curved axis) can be used to generate many different volumetric primitives (Dickinson *et al.*, 1991). Geon-based schemes are not ambiguous, but clearly not unique.

While geon representations have an intuitive appeal, the lack of quantitative information limits their applicability in environments where discrimination between qualitatively similar, but quantitatively different objects is performed. If an assembly line is making a 'family' of parts which differ only in scale, they would have identical geon representations, making discrimination between the different-sized items impossible without additional (quantitative) information (Flynn and Jain, 1991).

6.4   REPRESENTATION ISSUES IN IMAGE UNDERSTANDING

In this section we discuss object representation issues important for image understanding.

Among the volumetric schemes described in Section 6.3, decomposition schemes and geon-based ones are those mainly used in image understanding.

For instance, decomposition representations are used in application fields, such as robotics and biomedicine, where the approximation of an object or the reconstruction of object shapes from boundary points is required. In Section 6.5, we will see how object-based decomposition representations are built on data coming from different kinds of sensor data. In these applicative fields, the advantage provided by decomposition schemes is in having a volumetric representation which helps calculating integral properties and object space occupancy. Also space-based representations, essentially octrees, can be used in robotics, where an exact description of the object is not necessarily required (Samet, 1990a). Geons representations, instead, are suitable for image understanding applications because they model the human vision task.

More attention must be devoted to boundary schemes, since they are the most widely used for the recognition process. Object recognition is based, for the most part, on geometric features of the objects to be recognized, which include corners, edges and planar faces for polyhedra, as well as points, arcs of distinct curvature and regions of constant curvature for sculptured surfaces. Other features, such as axes of inertia, profile curves, surface textures properties, reflectance, etc. can also be employed. Thus, boundary representations, augmented with additional information, to help the matching process, are applied.

Often, not only geometric features enhancing the object recognition task are added to boundary representations, but also orientation, proximity, containment, covisibility, etc. Such relations are represented in the form of a graph or a hypergraph, since a hypergraph allows the encoding of relations involving more than two elements (see, for instance, Flynn and Jain (1991); Zhang *et al.* (1993)).

In what follows, we consider multiview representations, which describe objects through a finite set of viewer-centered descriptions. The approach consists of storing a set of projections (views) of all admissible 3D objects in a scene.

Up to now, we have considered different possible choices for object representation schemes. A fundamental issue in image understanding is whether to store the representation of an object in an object-centered or a viewer-centered coordinate frame. In the former case, we talk about an object-centered description of the object, in the latter, about a viewer-centered one. Note that only boundary or geon-based schemes are used in connection with a viewer-centered description.

Two well-known multiview representations are *characteristic views* (CVs) introduced by Freeman and Chakravarty in the early 80s (Freeman and Chakravarty, 1980) and *aspect graphs* (Koenderink and Van Doorn, 1979).

Given a 3D object, a *characteristic view domain* (CVD) is a contiguous open 3D subspace that contains all vantage points for which the perspective visible-line projections (views) have isomorphic line-junction graphs. A *line-junction graph* representing a given three-dimensional scene is a graph composed of all lines and vertices of a two-dimensional projection of the scene. The CVDs will be assumed to derive from the entire space exterior to the convex hull of the object.

Since all views corresponding to a CVD are related to each other through a linear transformation, we can select one view from a CVD, called the *Characteristic View* (or CV), to act as a representative view for that CVD. CVs corresponding to adjacent CVDs will have different line-junction graphs; however, this is not necessarily true for CVs corresponding to non-adjacent CVDs (Chen and Freeman, 1990). This approach makes it possible to represent a 3D object by a complete, finite set of characteristic views.

The recognition problem can then be efficiently solved by matching a given view of an unknown object with the set of characteristic views organized into a matching tree according to their line-junction graphs. Furthermore, this method can provide an efficient way for estimating orientation and pose of an object from images taken from arbitrary vantage points.

Since the concept of characteristic view was first introduced, the problem of computing all characteristic views of a given object was barely addressed. To simplify the problem, objects were limited to being planar-faced. However, even this problem was not solved until Wang and Freeman demonstrated the computation of the CVs for polyhedral objects (Wang and Freeman, 1990). In Wang and Freeman (1991), the concept of *canonical views*, i.e., the set of characteristic views which permits to represent an object without loosing essential information, was introduced and a method for extracting them was proposed. More recently, the *dominant views* of a solid object have been defined as the minimal subset of the object characteristic views that contain maximal visual information about the solid (Chen and Freeman, 1992). This allows to restrict the number of views to be stored.

Another possible solution for this problem, closely related to the CV approach, can be found in the definition of the *aspect graph*. Informally, an *aspect* can be regarded as the topological appearance of the object from a specific viewpoint, i.e., it is a specific view which is representative of the projections of the object from a connected set of viewpoints from which the object appears qualitatively similar (for example, identical line-junction graph, set of visible faces, etc.). In other words, an aspect graph captures all meaningful viewpoints of an object (Koenderink and Van Doorn, 1976).

Several definitions of aspect graphs have been proposed in the literature. However, since the aspect must capture the topology of an object, a correct definition should prevent the description of topologically different objects through the same aspect. For instance, the graphs depicted in Figure 6.4 are isomorphic, but the two objects are non-topologically equivalent. More formally, an *aspect* is a representation of the so-called *image structure graph* (Laurentini, 1994). Given a solid object and the corresponding line junction representation, the planar graph which has a node for each junction and an edge for each segment in such representation, is the *image structure graph*. Two aspects are said to be *topologically equivalent* if ordered adjacency and incidence

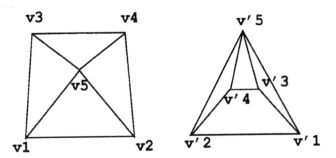

Figure 6.4 Examples of planar graphs representing aspects of non-topologically equivalent objects.

relations (as defined in Section 5.3.1) involving their vertices, edges, faces are the same in the two aspects.

To solve ambiguities frequently accepted in literature, in Laurentini (1994) a natural topological description of aspect has been proposed: for each polygonal face f of the graph, the list of vertices bounding f, sorted in clockwise order, is specified. An alternative is to maintain, for each node v of the graph, the adjacency lists (lists of nodes connected to v), sorted in clockwise order, as well as the list (sorted in clockwise order) of the vertices (nodes) composing the external polygonal face.

A change in the topology between aspects allows to represent them through a graph formalism. A *visual event* (or *transition*) is said to occur when the topology changes as the observer moves from one aspect to the other. Two aspects are *adjacent* if a visual event occurs by passing from one to the other one.

An *aspect graph* describing a solid object is thus a pair $G = (N, A)$ such that N is the set of nodes describing aspects and A is the set of arcs describing visual events. Figure 6.5 provides an example of an aspect graph.

There is no substantial difference between the CV method and the aspect graph: an aspect has the same definition as a characteristic view and the relations among all CVs of a given object can be obtained by simply recording the neighboring CVs during the computation of the CVDs. In other words, characteristic views correspond to nodes in an aspect graph. Moreover, for recognition purposes, the arcs of an aspect graph are generally not used.

Algorithms have been developed for computing aspect graphs (Gigus and Malik, 1990; Plantinga and Dyer, 1987; Kent *et al.*, 1986; Hansen and Henderson, 1989) for polyhedral objects. Many are characterized by a large time and space complexity (for a non-convex polyhedron with n sides and assuming orthographic projection, one existing algorithm requires $O(n^8)$ time and $O(n^8)$ storage to compute the aspect graph (Gigus and Malik, 1990)).

Some researches have taken a pragmatic approach to computing view-centered representations (Ikeuchi and Kanade, 1988; Jain and Hoffman,

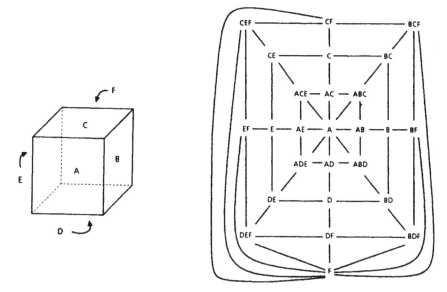

Figure 6.5 A cube and its aspect graph [reproduced from (Korn and Dyer, 1987), p. 100, copyright 1987, with kind permission from Elsevier Science].

1988; Korn and Dyer, 1987; Chen and Kak, 1989). Rather than obtaining an exact partition of the view sphere for each object, a discrete sampling of the view sphere is taken, images of the object are synthesized for each sample point and some appropriate features are computed from each image. In particular, in Hansen and Henderson (1989), a discrete approximation is obtained by placing a tessellated sphere around the model, where each of the polygons represents a different viewpoint. The tessellation can be arbitrarily fine, thus obtaining any desired granularity.

6.5   ALGORITHMS FOR SPATIAL REASONING

In the previous sections we have analysed models for describing solids, emphasizing the geometric modelling aspects. In this section, we show through examples, how geometric algorithms developed in the field of computational geometry have been employed for solving image understanding problems in application domains like robotics and biomedicine. One such problem is, for example, the reconstruction of a solid object from points or segments on its boundary, obtained through sensors. In more detail, we show how decomposition schemes, such as simplicial complexes, are often built by using suitable geometric techniques. This section does not have to be considered a survey of geometric methods applied to image understanding, but simply it intends to present some experiences which show the importance of geometric reasoning in image understanding problems.

In the examples we will describe below, a cell complex is used, in particular a special kind of two- and three-dimensional simplicial complex, i.e., the *Delaunay complex*, commonly called *Delaunay triangulation* in 2D and *Delaunay tetrahedralization* in 3D. Given an input set of points $\mathcal{P}$ in the plane (space), a Delaunay triangulation (tetrahedralization) of $\mathcal{P}$ is a triangulation (tetrahedralization) whose vertices are coincident with $\mathcal{P}$ and such that the circumcircle (circumsphere) of each triangle (tetrahedron) does not contain any point of $\mathcal{P}$ in its interior. Several algorithms have been proposed for the construction of a Delaunay triangulation and tetrahedralization (see Okabe *et al.* (1992) for a survey).

When input information is not a point set, but contains also a set of segments $\mathcal{S}$, it is desirable to built a Delaunay complex upon $\mathcal{P}$ in such a way that the segments in $\mathcal{S}$ are included in the edges of the complex.

Let $\mathcal{S}$ be a set of segments in the plane which do not intersect each other (except at their endpoints) and having their endpoints in $\mathcal{P}$. The pair $G = (\mathcal{P}, \mathcal{S})$ is called the *constraint graph*. A *triangulation* of $\mathcal{P}$ constrained by $\mathcal{S}$ is a triangulation of $\mathcal{P}$ containing the constraint graph as a subgraph. According to the definition proposed by Lee and Lin (1986), a *constrained Delaunay triangulation* of a point set $\mathcal{P}$ and a segment set $\mathcal{S}$ is a triangulation T of $\mathcal{P}$ constrained by $\mathcal{S}$ such that the circumcircle of each triangle t of T does not contain in its interior any other point of $\mathcal{P}$ which is *visible* from the three vertices of t (a point P_i is visible from P_j if the straight line joining P_i and P_j does not intersect any segment in $\mathcal{S}$). In Figure 6.6 a set of points $\mathcal{P}$ and segments $\mathcal{S}$ are depicted together with a constrained Delaunay triangulation of $\mathcal{P}$ and $\mathcal{S}$.

The inclusion of constraint segments may produce narrow triangles. An alternative way of embedding segments on a simplicial complex is to build a triangulation in which each input segment is represented as combination of edges of the simplicial complex. The idea consists in splitting the constraints into shorter segments by inserting new vertices on them, in such a way that the

Figure 6.6   A set of points and segments and the corresponding constrained Delaunay triangulation.

constrained Delaunay triangulation of the resulting set of subsegments will be a real Delaunay triangulation of the augmented vertex set.

If $G = (\mathcal{P}, \mathcal{S})$ is a constraint graph and $\mathcal{P}'$ is a point set containing $\mathcal{P}$, a triangulation $\mathcal{T}$ of $\mathcal{P}'$ *conforms* to G if every segment of $\mathcal{S}$ is union of edges of $\mathcal{T}$. A *conforming Delaunay triangulation* of $\mathcal{P}$ and $\mathcal{S}$ is a Delaunay triangulation of $\mathcal{P}'$ which conforms to G (Edelsbrunner and Tan, 1993). An example of conforming Delaunay triangulation is depicted in Figure 6.7.

In the literature several techniques have been developed for building constrained or conforming Delaunay triangulations with respect to a set of constraining segments. Algorithms for computing a constrained Delaunay triangulation have been proposed (Lee and Lin, 1986; Chew, 1989; Wang and Schubert, 1987; De Floriani and Puppo, 1992; Kao and Mount, 1991). The computation of a conforming Delaunay triangulation is generally a much harder problem. Constructing a conforming Delaunay triangulation of a constraint graph can force the introduction of a large number of points. At least $\Omega(mn)$ points need to be added to obtain a conforming Delaunay triangulation, where m is the number of constraint segments and n the number of input points. A common approach to produce a conforming Delaunay triangulation is to place sufficiently many points on the edges of the constraint graph (i.e., on the input segments), so that each interval has a circle that avoids all other edges (Faugeras *et al.*, 1990; Nackman and Srinivasan, 1991; Oloufa, 1991; Saalfeld, 1991). A drawback of such approach is that the number of points which must be added increases dramatically as the constraining segments are closer to each other: the total number of points at the end of the process can be at most $n(2m + 1)^m$. Edelsbrunner and Tan (1993) prove that $O(m^2 n)$ additional points (not necessarily upon the constraints) are sufficient to build a conforming Delaunay triangulation, and propose an algorithm that computes such points.

In three dimensions, it has been shown that the computation of a *constrained* Delaunay triangulation when the constraints are faces is an NP-complete problem (Ruppert and Seidel, 1989). The algorithm for computing a conforming Delaunay triangulation in 3D proposed by Faugeras *et al.* (1990) is not

Figure 6.7 A set of points and segments and the corresponding conforming Delaunay triangulation.

guaranteed to compute the mesh for an arbitrary partition of the constraint segments.

6.5.1   Object reconstruction from a set of points

In many applications, such as in biomedical problems, it is necessary to build a representation of an object from input data available in the form of a set of 3D points internal or on the boundary of the object. It is often desirable to have a representation of the whole boundary, which defines a valid object in an unambiguous way.

Given a set of points $\mathcal{P}$ in the 3D space, the general problem consists of computing a cell complex on $\mathcal{P}$ representing the object from which set $\mathcal{P}$ has been originated (for instance, through a laser range finder). Two possible descriptions of an object S can be provided: any 3D complex whose vertices are those of $\mathcal{P}$ defines a volumetric representation of S; any 2D complex, embedded in the space, whose vertices are formed by the subset of $\mathcal{P}$ representing the points on the surface of S provides a boundary description of S. The complex built on $\mathcal{P}$ must represent either a polyhedron or a closed surface of a polyhedral shape.

Point set $\mathcal{P}$ is either composed of points which lie on the boundary of the object, or belongs to cross sections obtained by intersecting the object with a collection of planes. When $\mathcal{P}$ consists of boundary points, trying all possible 2D cell complexes through a given set of n points is not feasible because of the combinatorial explosion of the number of possible solutions. Boissonnat (1982) proposed the use of the Delaunay tetrahedralization in connection with a heuristic to 'sculpture' a single connected shape. The strategy consists of two steps: first, a Delaunay tetrahedralization of the points of $\mathcal{P}$ is built, then, a sculpturing process is performed in order to obtain a tetrahedralization whose boundary is a collection of flat triangles containing as many points of $\mathcal{P}$ as possible (i.e., a triangulation which provides an approximation of the surface of the object).

The Delaunay tetrahedralization fills the interior of the convex hull of the points in $\mathcal{P}$ with tetrahedra. When there are points of $\mathcal{P}$ inside the convex hull, some tetrahedra must be eliminated until the points of $\mathcal{P}$ are on the boundary $\mathcal{B}$ of the resulting object. Such sculpturing process can be performed by eliminating one tetrahedron at a time in such a way that, at each step, the boundary of the current three-dimensional shape is a three-manifold polyhedral object. This is guaranteed when tetrahedra are eliminated by using the following rules:

1) every tetrahedron with three vertices on $\mathcal{B}$ can be eliminated;
2) if a tetrahedron not satisfying rule (1) has three faces on $\mathcal{B}$, it cannot be eliminated (otherwise a point $\mathcal{P}$ will be isolated);

3) if a tetrahedron not satisfying rule (1) has only one face on $\mathcal{B}$, it cannot be eliminated (otherwise $\mathcal{B}$ will intersect itself at one point $\mathcal{P}$);

4) if a tetrahedron not satisfying rule (1) has all its edges on $\mathcal{B}$, it cannot be eliminated (otherwise $\mathcal{B}$ will intersect itself along an edge joining two points of $\mathcal{P}$).

The algorithm stops when no more points of $\mathcal{P}$ can appear on $\mathcal{B}$ and no tetrahedron can be eliminated without decreasing the area of the boundary. After the sculpturing step has been performed, the object is represented by the set of the remaining tetrahedra. Boissonnat suggests an implementation of the reconstruction method whose complexity, in the worst case, is $O(n^2 \log n)$, where n is the number of input points. An example of the steps performed by the sculpturing process is provided in Figure 6.8.

A drawback of this algorithm is that the sculpturing process can get locked and it can stop before all innermost vertices have been included into the boundary. To overcome this problem, another approach, using a so-called *γ-neighborhood graph*, has been proposed in Veltkamp (1993). The γ-neighborhood graph is a structure connecting a set of points which includes the Delaunay tetrahedralization as a subgraph.

When an object is known through points belonging to cross sections obtained by intersecting it with a collection of planes, several *ad-hoc* methods have been proposed (see Boissonnat (1988) for a survey).

6.5.2 Object reconstruction from stereo data

Three-dimensional object reconstruction in robotics is often performed from stereo data, i.e., from data obtained by matching pairs of two-dimensional features extracted from at least two slightly different images. Stereo data are

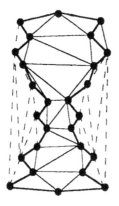 

Figure 6.8 An example of sculpturing process for a 2D object boundary reconstruction.

usually three-dimensional segments, which provide a kind of wireframe description of an object.

Hence, given a set of segments S in the 3D space, the general reconstruction problem consists of computing a cell complex representing an observed object, such that the segments of S lie on its boundary. Note that, in this application, the segments are always on the boundary of the object, since they correspond to pictorial evidences on its surface. Generally, when dealing with stereo data, the reconstruction of several objects, in a *scene*, rather than a single object is considered.

In this case also, a Delaunay tetrahedralization is used. Segments, and not only points, have to be constrained on the boundary of the tetrahedralization. Therefore, a Delaunay tetrahedralization is to be constructed in such a way that input segments are edges of tetrahedra.

Faugeras *et al.* (1990) proposed a method for generating a surface and free space description of an observed scene by first building a conforming Delaunay tetrahedralization on a set of stereo segments and, then, by marking those tetrahedra which occlude the stereo segments from a given view point: a tetrahedron is marked if it intersects the triangle having one vertex in the view point and a stereo segment as opposite edge. At the end of the process, the marked tetrahedra define the empty space between the scene and the view point, while the 2D simplicial complex formed by the boundary of the not-marked tetrahedra provide a surface description of the observed scene. Note that several view points can be considered during the empty tetrahedron detection. In Figure 6.9a the matched segments in the image plane of a trinocular stereo process are depicted, while Figures 6.9b and 6.9c show the cross section of a conforming Delaunay tetrahedralization with a plane parallel to the floor before and after the removal of the marked tetrahedra with respect to one view point.

The average performance of the construction of a conforming Delaunay tetrahedralization is equal to $O(m^3)$, if m is the number of input segments (Faugeras *et al.*, 1990). Nevertheless, the worst case time complexity can be exponential due to the number of points inserted on the constraint segments (Edelsbrunner and Tan, 1993).

An alternative two-dimensional approach has been developed for building a surface description of a scene, which is carried out in the two-dimensional space of the image plane, rather than in the three-dimensional space of the scene (Bruzzone *et al.*, 1992b). This method, which avoids errors due to camera calibration arising in the 3D reconstruction process, is suitable only for applications in which a volumetric description of the free space of the scene is not required. A two-dimensional constrained Delaunay triangulation is built on the 2D matched segments of the image plane of one of the cameras (Bruzzone *et al.*, 1992a). Then, the 2D simplicial complex in the 3D space is obtained by backprojecting each triangle from the image plane to 3D space: each vertex is pulled back on the corresponding 3D point (see Figure 6.10). All stereo segments appear in the final surface description: at each stereo segment

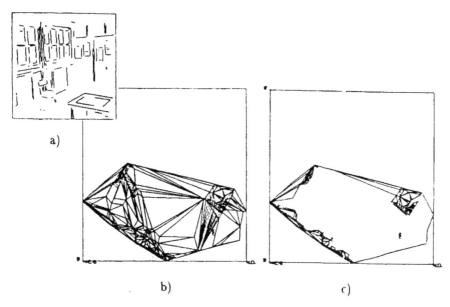

Figure 6.9 Matched segments in the image plane of a trinocular stereo process (a); cross section of a conforming Delaunay tetrahedralization before (b) and after (c) the marked tetrahedron removal [reproduced from Faugeras *et al.* (1990), pp. 69–70, copyright 1990, with kind permission from Elsevier Science].

corresponds a 2D segment, bounding a triangle of the 2D Delaunay triangulation in the image plane, which is backprojected on the corresponding stereo segment in 3D space. Therefore, the visibility constraint is automatically guaranteed, avoiding the heavy phase of marking tetrahedra representing free space, necessary when adopting a 3D approach.

The algorithm proposed in De Floriani and Puppo (1992) for the construction of a constrained Delaunay triangulation is used, which inserts each segment applying local modification to the topology of the current triangulation. The worst case complexity of the resulting reconstruction method is $O(m^3)$, where m is the number of input segments.

6.5.3 Free space map construction and path planning

Autonomous robot navigation is one of the most challenging problems in robotics. The main task in this field is to provide a robot with the capability of inspecting an unknown environment and of navigating through it. Data representing the environment are either stereo or ultrasound data. When the observed scene is described by 3D stereo segments, as discussed in the previous section, initial information about the position of the obstacles is achieved by projecting the segments and the view point on the plane on which a robot is

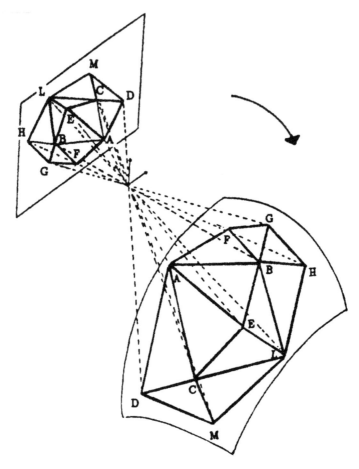

Figure 6.10 Perspective backprojection in the 3D space of the image triangulation [reproduced from Bruzzone *et al.* (1992b), p. 104, copyright 1992, with kind permission from Plenum Press].

considered to move. The segments which lie above the robot are not projected because they do not affect robot navigation. The free space map can be computed by applying geometric algorithms to the projected segments.

In Buffa (1993), a dynamic method for generating a free space map has been proposed. When new segments are projected into the plane, first a filtering process is performed in such a way to fuse those segments which are too close together (i.e., that correspond to the same feature, seen from two different view points). Then, a conforming Delaunay triangulation is computed on the remaining segments. In this case, not only dynamic insertions but also deletions of segments can occur, modifying the conforming triangulation as a consequence. The insertion and deletion of a segment is a quite expensive step, since adding a new segment requires the 'splitting' of that segment and, possibly, of other segments in its neighborhood; when eliminating a segment from the triangulation, the points inserted on it have also to be eliminated. A special structure, called *Delaunay tree*, is thus used to maintain the history of the triangulation process and make easier triangulation updating.

Free space is thus detected through visibility tests, which allow to identify which triangles represent free space. A ray r from the view point to a generic point P_i of the triangulation is considered. Starting from the view point all triangles intersecting r are marked, until a triangle bounded by an input segment is met. The polygon forming the border of the marked triangles represents the border between free space and the objects (obstacles) of the scene (see Figure 6.11).

An alternative method for exploring an unknown environment has been proposed in Bruzzone *et al.* (1993). After the projection process has been

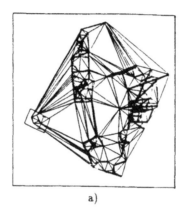

a)

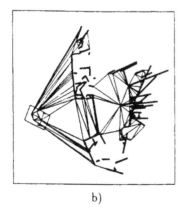

b)

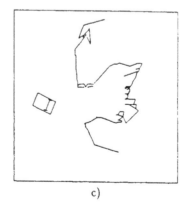

c)

Figure 6.11   An example of conforming Delaunay triangulation on a set of projected segments (a); triangles tessellating the free space (b); border between obstacles and free space (c) [reproduced from Buffa (1993)].

performed, the segments are connected together. Such method is based on the evaluation of a *local map*, which consists of a closed chain of segments representing the free space between the view point and the obstacles. The border of free space in front of the view point is represented by the open chain, mainly composed of those portions of 2D segments visible from the view point (see Figure 6.12a). Such chain of segments is called a *radial lower envelope* and can be computed in $O(m\alpha(m))$ in the worst case, where m is the number of segments and α is the inverse of the Ackermann's function. Segments in radial direction with respect to the view point are introduced to connect data, as well as fictitious segments representing potential (still unexplored) passages. A polygon is built by connecting the two extreme points of the open chain to the view point (see Figure 6.12b); then the actual local map is computed as union of the built polygon together with a square representing the robot (see Figure 6.12c).

The global map of the free space is a multiply connected polygon, with several internal contours (i.e., holes) and an external one, obtained by iteratively merging a current with a local map represented as a simply connected polygon.

Robot path planning is thus performed on the final free space map by subdividing the map into regions, defined by a *Voronoi diagram* built on the sides of the global map. A *Voronoi region* of a given segment s represents that area of the free space closer to s than to any other segment (Drysdale, 1979). Any path following the edges of the Voronoi diagram defines a free-collision trajectory.

In Buffa (1993), the robot motion is planned by connecting the baricenters of two adjacent triangles of the free space, generating a path from the starting point to the goal: the robot moves on the edges of a Voronoi diagram built on the vertices of the triangles of the free space. Note that in such a way the robot is not guaranteed to be always equidistant from the obstacles.

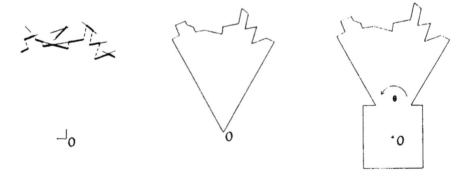

Figure 6.12   Radial lower envelope of a set of segments (a); local map (b); real local map with robot encumbrance (c) [reproduced from Bruzzone *et al.* (1993), copyright 1993, with the permission of World Scientific Publishing].

6.6 CONCLUSIONS

In the paper, we have considered the problem of modelling solid objects and we have discussed the requirements that a representation must satisfy for recognition purposes. The state of the art on object representation schemes and multiview representations (which provide an object description as a collection of objects views) has been presented with a special emphasis on representation schemes for image understanding.

We have also shown, through some examples, how techniques developed in the computational geometry literature can be successfully applied to solve well-known problems in image understanding and robotics, namely the object and scene reconstruction problem and the free space reconstruction for robot navigation. In particular, we have shown the wide applicability of geometric structures, like the Delaunay triangulation in two or three dimensions, and of the corresponding construction algorithms.

REFERENCES

Ansaldi, S., De Floriani, L. and Falcidieno, B. (1985) Geometric modeling of solid objects by using a face adjacency graph representation. *Computer Graphics* **19**, 3, 131–139.

Baumgart, B.G. (1972) Winged-edge polyhedron representation. *Technical Report STAN-CS-320*, Computer Science Department, Stanford University, Stanford, California.

Besl, P.J. and Jain, R.C. (1985) Three-dimensional object representation. *ACM Comput. Surveys* **17**, 1, 75–145.

Besl, P.J. (1988) Geometric modelling and computer vision. *Proceedings IEEE* **76**, 8, 936–958.

Biederman, I. (1985) Human image understanding: Recent research and a theory. *Computer Vision, Graphics and Image Processing* **32**, 29–73.

Biederman, I. (1987) Recognition-by-components: a theory of human image understanding. *Psychol. Rev.* **94**, 2, 115–147.

Binford, T.O. (1984) Survey of model-based image analysis systems. *ACM Trans. on Graphics* **3**, 4, 266–286.

Boissonnat, J.D. (1982) Geometric structures for three-dimensional shape representation. *Int. J. Robotics Research* **1**, 1.

Boissonnat, J.D. (1988) Shape reconstruction from planar cross-sections. *Computer Vision, Graphics and Image Processing* **44**, 1–29.

Brady, J., Nandhakumar, N. and Aggarwal, J. (1988) Recent progress in the recognition of objects from range data. *Proc. 9th Int. Conf. Pattern Recognition*, 85–92.

Brisson, E. (1989) Representing geometric structures in *d*-dimensions: Topology and order. *Proc. 5th ACM Symp. on Computational Geometry*, 218–227, Saarbruecken, Germany.

Brooks, R.A. (1981) Symbolic reasoning among 3-D models and 2-D images. *Artificial Intelligence* **17**, 285–348.

Bruzzone, E., Cazzanti, M., De Floriani, L. and Mangili, F. (1992a) Applying two-dimensional Delaunay triangulation to stereo data integration. *Proc. European Conf. on Computer Vision*, 368–372, S. Margherita, Italy.

Bruzzone, E., Garibotto, G. and Mangili, F. (1992b). Three-dimensional surface reconstruction using Delaunay triangulation in the image plane. In: *Visual form – analysis and recognition* (eds C. Arcelli, L.P. Cordella and G. Sanniti di Baja), 99–108. Plenum Press, New York.

Bruzzone, E., De Floriani, L., Gallo, P. and Mangili, F. (1993). A real-time stereo vision system for a mobile robot with exploratory tasks. *Proc. 7th Int. Conf. Image Analysis and Processing*, 591–598, Bari, Italy.

Buffa, M. (1993) *Robot navigation by using stereo vision and Delaunay triangulation*. Ph.D. Thesis, University of Nice (in French).

Chen, C. and Kak, A. (1989) A robot vision system for recognizing 3-D objects in low-order polynomial time. *IEEE Trans. Syst. Man Cybern.* **19**, 6, 1535–1563.

Chen, S. and Freeman, H. (1990) Computing characteristic views of quadric-surfaced solids. *Proc. 10th Int. Conf. on Pattern Recognition*, Atlantic City, New Jersey.

Chen, S. and Freeman, H. (1992) The dominant views of solid objects. *Proc. 11th Int. Conf Pattern Recognition*, The Hague, The Netherlands.

Chew, L.P. (1989) Constrained Delaunay triangulations. *Algorithmica* **4**, 97–108.

Chin, R.T. and Dyer, C.R. (1986) Model-based recognition in robot vision. *ACM Comput. Surveys* **18**, 1, 67–108.

De Floriani, L. and Puppo, E. (1992) An on-line algorithm for constrained Delaunay triangulation. *CVGIP: Graphical Models and Image Processing* **54**, 3, 290–300.

Dickinson, S.J., Rosenfeld, A. and Pentland, A.P. (1991) Primitive-based shape modeling and recognition. *Proc. Workshop on Visual Form*, 213–229, Capri, Italy.

Dobkin, D. and Laszlo, M. (1987) Primitives for the manipulation of three-dimensional subdivisions. *Proc. 3rd ACM Symp. on Computational Geometry Models*, 86–99, Canada.

Drysdale, R.L. (1979) *Generalized Voronoi diagrams and geometric searching*. Ph.D Thesis, Dept. Comp. Sci., Stanford University, Stanford, California.

Edelsbrunner, H. (1987) *Algorithms in combinatorial geometry*. Springer Verlag, Berlin, Germany.

Edelsbrunner, H. and Tan, T.S. (1993) An upper bound for conforming Delaunay triangulations. *Discrete Comput. Geom.* **10**, 197–213.

Faugeras, O.D., Le Bras-Mehlman, E. and Boissonnat, J.D. (1990) Representing stereo data with the Delaunay triangulation. *Artificial Intelligence* **44**, 41–89.

Ferrucci, V. and Paoluzzi, A. (1991) Extrusion and boundary evaluation for multi-dimensional polyhedra. *Computer Aided Design* **23**, 1, 40–50.

Flynn, P.J. and Jain, A.K. (1991) CAD-based computer vision: from CAD models to relational graphs. *IEEE Trans on Patt. Analysis and Mach. Intell.* **PAMI 13**, 2, 114–132.

Freeman, H. and Chakravarty, I. (1980) The use of characteristic views in the recognition of three-dimensional objects. In: *Pattern recognition in practice* (eds E. Gelsema and L. Kanal). North-Holland, Amsterdam, The Netherlands.

Gigus, Z. and Malik, J. (1990) Computing the aspect graph for line drawings of polyhedral objects. *IEEE Trans. on Patt. Analysis and Mach. Intell.* **PAMI 12**, 2, 113–122.

Guibas, L. and Stolfi, J. (1985) Primitives for the manipulation of general subdivisions and computation of Voronoi diagrams. *ACM Trans. on Graphics* **4**, 74–123.

Gursoz, E.L., Choi, Y. and Prinz, F.B. (1990) Vertex-based representation of non-manifold boundaries. In: *Geometric modeling for product engineering* (eds M.J. Wozny,

J.U. Turner and K. Preiss), 107–130. Elsevier North Holland, Amsterdam, The Netherlands.

Hansen, C. and Henderson, T.C. (1989) CAD-based computer vision. *IEEE Trans. on Patt. Analysis and Mach. Int.* **PAMI-11**, 11, 1181–1193.

Ikeuchi, K. and Kanade, T. (1988) Automatic generation of object recognition programs. *Proceedings IEEE* **76**, 8, 1016–1035.

Jackins, C.L. and Tanimoto, S.L. (1984) Octrees and their use in representing three-dimensional objects. In: *Solid modeling by computers: from theory to applications.* Plenum Press, New York.

Jain, A.K. and Hoffman, R.L. (1988) Evidence-based recognition of 3-D objects. *IEEE Trans. on Patt. Analysis and Mach. Intell.* **PAMI-10**, 6, 783–802.

Kao, T.C. and Mount, D.M. (1991) Incremental construction and dynamic maintenance of constrained Delaunay triangulations. *Proc. 3rd Canadian Conf. on Computational Geometry*, 170–175.

Kent, E.W., Schneir, O. and Hong, T.H. (1986) Building representations from fusions of multiple views. *Proc. IEEE Conf. Robotics and Automation*, 1634–1639, San Francisco, California.

Koenderink, J.J. and van Doorn, A.J. (1976) The singularities of the visual mapping. *Biological Cybernetics* **24**, 51–59.

Koenderink, J.J. and van Doorn, A.J. (1979) The internal representation of solid shape with respect to vision. *Biol. Cybern.* **32**, 211–216.

Korn, M.R. and Dyer, C.R. (1987) 3-D multiview object representations for model-based recognition. *Pattern Recognition* **20**, 1, 91–103.

Laurentini, A. (1994) *Topological recognition of polyhedral objects from multiple views.* Technical Report, Polytechnic of Torino, Italy.

Lee, D.T. and Lin, A.K. (1986) Generalized Delaunay triangulation for planar graphs. *Discrete Comput. Geom.* **1**, 201–217.

Levine, M.D., Bergevin, R. and Nguyen, Q.L. (1991) Shape description using geons as 3D primitives. *Proc. Workshop on Visual Form*, 363–377. Capri, Italy.

Lienhardt, P. (1991) Topological models for boundary representations: a comparison with *n*-dimensional generalized maps. *Computer Aided Design* **23**, 1, 59–82.

Mäntylä, M. (1987) *An introduction to solid modeling.* Computer Science Press, Rockville, Maryland.

Marr, D. and Nishihara, K.H. (1978) Representation and recognition of the spatial organization of three-dimensional shape. *Proc. Royal Society of London* **B200**, 269–294.

Murabata, S. and Higashi, M. (1990) Non-manifold geometric modeling for set operations and surface operations. *Proc. IFIP 5.2 Workshop on Geometric Modeling*, Rensselaerville, New York.

Nackman, L.R. and Srinivasan, V. (1991) Point placement for Delaunay triangulation of polygonal domains. *Proc. 3rd Canadian Conf. on Computational Geometry*, 37–40.

Okabe, A., Boots, B. and Sugihara, K. (1992) *Spatial tessellations – concepts and applications of Voronoi diagrams.* John Wiley, New York.

Oloufa, A.A. (1991) Triangulation applications in volume calculation. *Journal Comput. Civil Engineering* **5**, 103–119.

Plantinga, H. and Dyer, C. (1987) *The aspect representation.* Technical Report CSTR-683, Dep. Comput. Sci., University Wisconsin, Madison, Wisconsin.

Requicha, A.A.G. (1980) Representations of rigid solids: Theory, methods, and systems. *ACM Computing Surveys* **12**, 4, 437–464.

Requicha, A.A.G. and Voelcker, H.B. (1982) Solid modeling: a historical summary and contemporary assessment. *IEEE Computer Graphics and Applications* **2**, 2, 9–24.

Requicha, A.A.G. and Voelcker, H.B. (1983) Solid modeling: current status and research directions. *IEEE Computer Graphics and Applications* **3**, 7, 25–37.

Rossignac, J.R. (1991) Through the cracks of the solid modeling milestone. In: *Eurographics 91 state of the art report*, 23–109.

Ruppert, J. and Seidel, R. (1989) On the difficulty of tetrahedralizing 3-dimensional non-convex polyhedra. *Proc. 5th ACM Symp. on Computational Geometry*, 380–393.

Saalfeld, A. (1991) Delaunay edge refinements. *Proc. 3rd Canadian Conf. on Computational Geometry*, 33–36.

Samet, H. (1990a) *The design and analysis of spatial data structures*. Addison-Wesley, Reading, Massachusetts.

Samet, H. (1990b). *Applications of spatial data structures*. Addison Wesley, Reading, Massachusetts.

Tamminen, M. (1984) Comment on quad- and octrees. *Communications of the ACM* **27**, 3, 248–249.

Veltkamp, R.C. (1993) 3D computational morphology. *EURO-GRAPHICS '93*, 115–127.

Wang, C.A. and Schubert, L. (1987) An optimal algorithm for constructing the Delaunay triangulation of a set of line segments. *Proc. 3rd ACM Symp. on Computational Geometry*, 223–232.

Wang, R. and Freeman, H. (1990) Object recognition based on characteristic view classes. *Proceedings 10th Int. Conf. on Pattern Recognition*, Atlantic City, New Jersey.

Wang, R. and Freeman, H. (1991) From characteristic views to canonical views. *Proc. 6th Int. Conf. on Image Analysis and Processing*, Como, Italy.

Weiler, K. (1985) Edge-based data structures for solid modeling in a curved-surface environment. *IEEE Computer Graphics and Applications* **5**, 1, 21–40.

Weiler, K. (1986) *Topological structures for geometric modeling*. Ph.D. dissertation, Dept. Computer and System Engineering, Rensselaer Polytechnic Inst., Troy, New York.

Woo, T.C. (1985) A combinatorial analysis of boundary data structures schemata. *IEEE Computer Graphics and Applications* **5**, 3, 19–27.

Zhang, S., Sullivan, G.D. and Baker, K.D. (1993) The automatic construction of a view-independent relational model for 3-D object recognition. *IEEE Trans. on Patt. Analysis and Mach. Int.* **PAMI-15**, 6, 531–544.

7

Vision as Uncertain Knowledge

7.1 INTRODUCTION TO VISION AS KNOWLEDGE

Visual tasks require a considerable amount of information to be acquired, formatted, indexed and processed in order to achieve acceptable results in real environments. In the recent years – much more than in the past – attention has been devoted to integrating knowledge elements into sound, robust formal schemes, in the attempt to capture the interactions occurring in a real visual problem among the scene objects, acquisition system, observer and *a priori* information.

Knowledge-based (*model-based*) *vision* (Binford, 1982; Brooks, 1983) is an

ARTIFICIAL VISION
ISBN 0-12-444816-X

Copyright © 1997 Academic Press Ltd
All rights of reproduction in any form reserved

approach that offers successful solutions, by mainly concentrating on *visual knowledge representation* (Reiter and Mackworth, 1989/90) as a key issue in the design of machine vision systems.

The real world is inherently dynamical, ambiguous and affected by casual phenomena. Additional uncertainty occurs, since reconstructing shapes and spatial relations from visual data is an inverse, ill-posed problem, that admits infinitely many solutions (for a review, see Verri, 1996). Therefore, visual knowledge is incomplete, inexact and uncertain, and representation techniques should consistently account for these issues.

Several techniques have been proposed and applied, which may be grouped (Pearl, 1988) into *extensional* (*syntactic*) and *intensional* (*semantic*). The former are largely based on logic formalisms, which are extended by including truth assignments to phrases and mechanisms to combine them. The latter are based on probabilistic formalisms to assign certainty measures to elements of knowledge. In spite of their relatively higher computational complexity, intensional formalisms appear more adequate to represent visual problems, because they provide clear, consistent representations, which, at the same time, are flexible enough to fit to dynamic worlds.

The present paper provides an introductory review to probabilistic (or Bayes) networks (Pearl, 1986, 1988) – an intensional formalism that offers a consistent treatment of uncertain knowledge – and its applications to visual domains.

The paper is divided into two main parts, the first one dealing with the basics of Bayes networks and inference mechanisms, the second with its applications. In particular, Section 7.2 introduces the basic formalism and Section 7.3 extends the previous discussion to include a formal treatment of planning functionalities. Section 7.4 discusses how objects and relations typical of visual domains can be represented by Bayes nets, while the next section deals with the complex role of nets for visual inference. Planning visual actions is discussed in Section 7.6. Our conclusions are reported in Section 7.7.

7.2 PROBABILISTIC NETWORKS BASICS

We recall the basic definitions and properties of probabilistic networks, by emphasizing those directly related to vision; for a deeper insight into the formalism, the reader has at his or her disposal excellent textbooks and articles (Pearl 1986, 1988; Neapolitan, 1990; Charniak, 1991).

7.2.1 A view of the universe

A knowledge representation scheme is grounded on an *ontology*, a set of elements (entities, relations) and inference mechanisms assumed to exist. In our case, the universe is described – as in a stochastic process – by a multivariate discrete probability distribution. Entities are casual variables, originating evidences, beliefs, hypotheses; relations are dependencies among casual variables in form of conditional probabilities. A network model is adopted as a basic scheme to arrange entities (node labels) and relations (arc labels).

As an example, let us consider the simple net in Figure 7.1. The causal variable H is associated to a set of hypotheses (labels, decisions); E is associated to a set of evidences (data, symptoms). $P(H)$ and $P(E)$ are *a priori* known probabilities of occurrence of a hypothesis and evidence respectively. Both are typically in the form of vectors:

$$P(H) = (p(H = h_1), p(H = h_2), \ldots, p(H = h_n));$$

n number of plausible hypotheses;

$$P(E) = (p(E = e_1), p(E = e_2), \ldots, p(E = e_m));$$

m number of evidences.

$P(E|H)$ is a matrix encoding the set of *a priori* probabilities that the possible evidences follow from a verified hypothesis H.

The given elements of knowledge can be used in the opposite sense, to infer the plausibility of a hypothesis element, given an observed piece of evidence. The key point is the *Bayes rule*:

$$p(H = h_j | E = e_i) = \frac{p(E = e_i | H = h_j) p(H = h_j)}{p(E = e_i)}$$

Therefore, *a priori* and *a posteriori* (i.e., after an observation) statements about the world are associated in a sort of *antecedent-consequent pair*, as in ordinary inference rules.

The associative steps can be combined into *reasoning chains*, thus propagating evidences through the net starting by direct observations. Both the network and the inference mechanism will be formally presented in the next section.

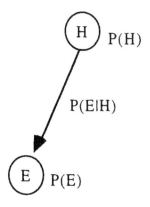

Figure 7.1

7.2.2   Formal issues

A probabilistic (Bayes) network is a pair:

$$\{G(N,E),M\}$$

with:

$G(N,E)$: Directed Acyclic Graph (DAG);

N: set of nodes, labelled with casual variables X_i;

E: set of directed arcs, i.e., ordered pairs $e_{ij} = (n_j, n_i)$

and:

M: set of conditional probability distributions

$$M \equiv \{P(X_i|S(X_i)) : X_i \in W\}$$

$S(X_i)$: set of father nodes of X_i

$$S(X_i) \equiv \{X_j \in W : i \neq j, \exists\, e_{ij} = (n_j, n_i) \in E\}.$$

The definition of *conditional probability* is here recalled:

$$P(X_i|S(X_i)) \equiv \{p(X_i = x_j^i \mid Y_1 = y_{k_1}^1, Y_2 = y_{k_2}^2, \ldots, Y_m = y_{k_m}^m) : X_i \in \Omega;$$

$$Y_1, Y_2, \ldots, Y_m \in S(X_i)); j = 1, \ldots, t_i;$$

$$k_1 \in \{1, \ldots, t_1\}, \ldots, k_m \in \{1, \ldots, t_m\}\}.$$

The variables X_i and X_j are said to be *conditionally independent given the variable* X_k iff:

$$\forall p \in \{1, \ldots, t_i\}\ \forall q \in \{1, \ldots, t_j\}\ \forall r \in \{1, \ldots, t_k\}$$

$$p(X_i = x_i^p \mid X_j = x_j^q, X_k = x_k^r) = p(X_i = x_i^p \mid X_k = x_k^r). \tag{7.1}$$

The topology of a net immediately reflects the latter concept: the nodes labelled with X_i and X_j are not directly connected, but are separated by the common father node X_k.

Completeness and consistency   A Bayes net is a *complete* and *consistent* representation of a multivariate probability distribution.

Given an L-variate discrete probability distribution $P(X_1, X_2, \ldots, X_L)$, completeness states that to any possible L-ple $(x_1, \ldots, x_L)$ from $(X_1, X_2, \ldots, X_L)$ is associate a probability $p(x_1, \ldots, x_L)$.

The consistency conditions are:

$$\forall x_1, \ldots, x_L\ \ 0 \leqslant p(x_1, \ldots, x_L) \leqslant 1$$

$$\sum_{x_1, \ldots, x_L} p(x_1, \ldots, x_L) = 1. \tag{7.2}$$

7.2.3 Inference and its multiple sides

Given $\mathfrak{I}$, the set of collected data/evidences, the *belief vectors* associated to the casual variables are defined as follows:

$$\text{Bel}(x_k^i) \equiv p(X_i = x_k^i | \mathfrak{I})$$

$$\text{Bel}(X_i) \equiv (p(X_i = x_1^i | \mathfrak{I}), p(X_i = x_2^i | \mathfrak{I}), \ldots, p(X_i = x_{t_i}^i | \mathfrak{I})).$$

If the network is a tree (see Figure 7.2):

$$\mathfrak{I} = \zeta_x^+ \cup \zeta_x^-$$

where:

$$\zeta_x^- \equiv \zeta_x \cup \zeta_x^{1-} \cup \zeta_x^{2-} \cup \cdots \cup \zeta_x^{m-}.$$

Now we calculate the belief values at a node with simple manipulations:

$$\text{Bel}(X_i) = p(X_i | \zeta_x^+, \zeta_x^-)$$

$$= \frac{p(X_i, \zeta_x^+, \zeta_x^-)}{p(\zeta_x^+, \zeta_x^-)} \cdot \frac{p(X_i, \zeta_x^+)}{p(X_i, \zeta_x^+)} \cdot \frac{p(X_i, \zeta_x^+)}{p(X_i, \zeta_x^+)}$$

$$= \alpha p(\zeta_x^- | X_i) p(X_i | \zeta_x^+)$$

with α a normalization constant.

By defining:

$$\pi(x_i) \equiv p(x_i | \zeta_X^+); \qquad \lambda(x_i) \equiv p(\zeta_X^- | x_i)$$

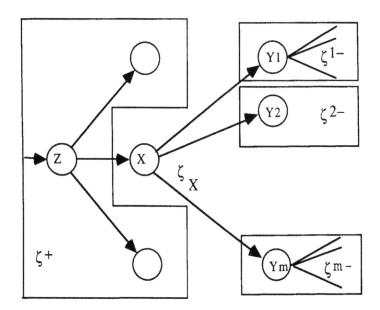

Figure 7.2

and the corresponding vectors:

$$\lambda(X) \equiv (\lambda(x_1), \ldots, \lambda(x_n))$$

$$\pi(X) \equiv (\pi(x_1), \ldots, \pi(x_n)).$$

The basic equation writes:

$$\text{Bel}(X) = \alpha\lambda(X)\pi(X). \tag{7.3}$$

Equation (7.3) describes the *fusion mechanism* operating at a node, on which are grounded all the inference processes on the network. Several aspects of inference emerge from this fundamental result. Inference information appears as a process of *fusion of information distributed at the nodes*. It consists in updating the belief values (*belief revision*): belief values can either be increased (reinforced) or lowered, thereby defining *non-monotonic* reasoning steps. Moreover, inference is inherently dynamical, in the sense that the belief values are transmitted through the links in both bottom-up and top-down ways.

In fact, two contributions determine the belief values at the node X: one, received from the son nodes of X, called the *diagnostic support* $\lambda(X)$. The other term in (7.3), $\pi(X)$, is called the *causal support* to X and comes from its father node, which collects causal contributions from its ancestors and diagnostic terms from the other sons, siblings of X. The root node only receives diagnostic support from its sons.

Studying the inference scheme in probabilistic networks is an open research field in computer science and artificial intelligence. The general problem of inference is NP-complete (Cooper, 1990); a number of researchers have investigated the properties of specific inference algorithms and their role in automated reasoning (Pearl, 1987a, 1987b; Henrion, 1988; Schachter and Peot, 1990; Bhatnagar and Kanal, 1993; Dagum and Chavez, 1993; Peot and Schachter, 1993). Results are also available concerning languages and software environments to define Bayes nets and realize automated reasoning (Breese, 1992; Jensen *et al.*, 1992; Goldman and Charniak, 1993).

Even though we have not given all the details necessary to design a specific inference algorithm, we shall anyway outline how it works when the probabilistic net is a tree.

7.2.4 The basic fusion algorithm

The basic fusion algorithm can be represented:

$$\forall X \in \Omega$$

REPEAT:

$$\text{Belief revision}$$

$$\lambda(x_i) = \prod_{k=1}^{m} \lambda_{Y_k}(x_i) \; \forall i \in \{1, \ldots, n\} \qquad \text{updating vec. } \lambda(X)$$

$$\pi(x_i) = \beta \sum_j p(x_i \mid z_j)\pi_x(z_j) \; \forall i \in \{1, \ldots, n\}; \qquad \text{updating vec. } \pi(X)$$

$$\text{Bel}(X) = \alpha\lambda(X)\pi(X); \qquad \text{updating vector Bel}(X)$$

Bottom-up propagation

$$\lambda_X(z_j) = \sum_{i=1}^{n} p(x_i \mid z_j)\lambda(x_i) \; \forall j; \qquad X \text{ transmits } \lambda_X(Z) \text{ to father } Z$$

Top-down propagation

$$\pi_{Y_k}(x_i) = \alpha' \frac{\text{Bel}(x_i)}{\lambda_{Y_k}(x_i)}; \qquad X \text{ transmits } \pi_{Y_k}(x_i) \text{ to } k\text{th son } Y_k$$

UNTIL: $\text{Bel}_t(X) = \text{Bel}_{t-1}(X)$.

7.3 PLANNING IN UNCERTAIN UNIVERSES

The basic network topology and the related inference scheme can be extended to include *learning* (Pearl, 1988; Sucar and Gillies, 1994) and *planning* (Dean and Wellman, 1991) mechanisms, thus providing a comprehensive knowledge representation framework. In the following we shall concentrate on the latter topic, since an extensive research work has been carried out on it.

7.3.1 Planning as a decision problem

Given a goal and a set of executable actions, the aim of *planning* is providing an ordered sequence of actions to achieve that goal. Planning under uncertain conditions can be formulated in terms of Bayesian models, complemented with concepts from decision theory and information theory.

Let $\{G(N, E), M\}$ be a Bayes net, and $X_d \in W$ a random variable associated to the goal, whose values represent mutually exclusive hypotheses x_k^d $(k = 0, \ldots, t_d - 1)$, which we shall call *decisions* $d_k \in \{d_0, d_1, \ldots, d_{t_d-1}\}$ in the present context.

The aim of planning is *instantiating* X_d with x_0^d representing the most plausible hypothesis – or the optimal decision – on the basis of a set of evidences $\Im$.

In order to solve the problem, the system may perform one or more *information gathering actions* α_{e_i}, belonging to a pre-defined set A, and obeying to specified preconditions. The outcome of an action is instantiating a network variable with a specific evidence $X_{e_i} = x_k^{e_i}$, which, in turn, triggers an inference procedure through the net.

In uncertain universes it is impossible to guarantee, in general, that a plan can reach a goal. We are forced to introduce the concept of *partial satisfiability*: a goal has been reached whenever the uncertainty (e.g., the list of beliefs about a variable) is kept to a minimum.

Given a set of actions A, selecting the optimal action at a given time requires evaluating the usefulness of all possible sequences of executable actions: if $(\alpha_{e_1}, \ldots, \alpha_{e_n})$ has been selected, then the optimal action is obviously α_{e_1}. Being such an intractable approach, approximate methods have been proposed, still based upon measures $V(\alpha_{e_i})$ of utility (value) of an evidence provided by an action α_{e_i}.

7.3.2 Estimating value and goodness

Expected value of sample information In decision theory, the optimal hypothesis/decision d^* is given by:

$$d^* \equiv \arg\max_{d_i} \mathrm{EV}(d_i)$$

where

$$\mathrm{EV}(d_i) \equiv \sum_{j=1}^{t_d} U(d_i, x_j^d) \cdot \mathrm{Bel}(x_j^d)$$

The *payoff matrix* $U(t_d \times t_d)$ expresses how good is the hypothesis/decision d_i, given that the verified hypothesis is x_j^d. Then, $\mathrm{EV}(d_i)$ (Expected Value of d_i) is a weighted sum of the belief values $\mathrm{Bel}(x_j^d)$ of the goal variable X_d with the goodness values associated to d_i. On this basis, we can estimate the usefulness of an action α_e, through the Expected Value of Sample Information (EVSI).

$$V_{\mathrm{EVSI}}(\alpha_e) \equiv \mathrm{EV}_{\alpha_e} - \mathrm{EV}_0, \text{ where } \mathrm{EV}_0 \equiv \mathrm{EV}(d^*) \tag{7.4}$$

and EV_{α_e} is defined as:

$$\mathrm{EV}_{\alpha_e} \equiv \sum_{k=1}^{t_e} \left[\max_{i \in \{1, \ldots, n\}} H(d_i, x_k^e) \right] \cdot \mathrm{Bel}(x_k^e)$$

where

$$H(d_i, x_k^e) \equiv \sum_{j=1}^{t_d} U(d_i, x_j^d) \cdot \mathrm{Bel}(x_j^d)|_{X_e = x_k^e}$$

where $\mathrm{Bel}(x_j^d)|_{X_e = x_k^e}$ is the updated value of $\mathrm{Bel}(x_j^d)$ after instantiating the variable X_e with x_k^e. Therefore, the value $V_{\mathrm{EVSI}}(\alpha_e)$ of the information yielded by an action α_e is the difference between the expected value EV_{α_e} associated to the optimal decision $d_{\alpha_e}^*$ *after hypothetically executing* α_e, and the expected value EV_0 associated to the optimal decision d^*, calculated *without the information arising from executing* α_e. Then, α_e is useful iff $V_{\mathrm{EVSI}}(\alpha_e) > 0$. We also observe that, to calculate $V_{\text{EVSI}}(\alpha_e)$ we should apply t_e times the probabilistic inference algorithm to compute $\text{Bel}(x_j^d)$ for each possible instance of X_e.

The Shannon measure An action α_e can be thought as an information source whose value can be estimated – as in the information theory – by its effects on the probability distribution of a goal variable X_d. A well-known estimate of uncertainty encoded in the probability distribution of X_d is its *entropy*:

$$\text{ENT}(X_d) \equiv -\sum_{i=1}^{t_d} p(x_i^d) \log p(x_i^d).$$

In the same way the entropy of a conditioned distribution $P(X_d | X_e)$ is:

$$\text{ENT}(X_d | X_e) \equiv -\sum_{i=1}^{t_e} \sum_{j=1}^{t_d} p(x_j^d | x_i^e) \log p(x_j^d | x_i^e).$$

By subtracting from $\text{ENT}(X_d | X_e)$ the prior estimate of uncertainty on X_d – $\text{ENT}(X_d)$ – we get an estimate of how much the former has been reduced by 'consulting' the information source X_e: It is the *Shannon measure*.

$$I(X_d, X_e) \equiv \text{ENT}(X_d | X_e) - \text{ENT}(X_d)$$

$$= -\sum_{i=1}^{t_e} \sum_{j=1}^{t_d} p(x_j^d | x_i^e) \log \frac{p(x_j^d | x_i^e)}{p(x_j^d) p(x_i^e)}.$$

The 'most valuable' information source α_e^*, is the one that minimizes $I(X_d, X_e)$:

$$\alpha_e^* \text{ with } e \equiv \arg\min_i I(X_d, X_i).$$

Goodness and strategies A more complex model of utility of an action descends from balancing its value and cost in terms of a *goodness function* $G(\alpha_{e_i})$, which in practical problems is based on heuristics.

The simplest choices are:

$$G(\alpha_{e_i}) \equiv V_{\text{EVSI}}(\alpha_{e_i}) - C(\alpha_{e_i});$$

$$G(\alpha_{e_i}) \equiv -\frac{V_{\text{I}}(\alpha_{e_i})}{C(\alpha_{e_i})};$$

where $V_{\text{EVSI}}(\alpha_{e_i})$ and $V_{\text{I}}(\alpha_{e_i})$ are the values of information produced by α_{e_i}, while $C(\alpha_{e_i})$ is its cost. The most promising action is then:

$$\alpha^* \equiv \arg\max_{\alpha_{e_i} \in A} G(\alpha_{e_i})$$

where A is the set of all executable actions.

The latter technique defines a *myopic planning strategy*, since is based on the estimate of the impact of one single action (i.e., the next one). A more adequate nonmyopic strategy should take into account sequences of forthcoming actions.

Several solutions have been proposed in the literature (Dean and Wellman, 1991; Heckerman *et al.*, 1993; Rimey and Brown, 1994). The planning algorithm presented in the next subsection implements one such strategy.

7.3.3 A nonmyopic algorithm

A nonmyopic planning algorithm is given below:

$$\mathfrak{I} = \varnothing$$

LOOP:

    1 – Compute the most plausible value of goal variable:

$$d^* = \arg\max_{d_i} EV(d_i).$$

    2 – Test whether further actions are worthwhile:

        IF $\forall \alpha_{e_i} G(\alpha_{e_i}) \leqslant 0$:

        – compute $V_{s_k} - C_{s_k}$ for sequence s_k selected heuristically

        – IF $\forall s_k V_{s_k} - C_{s_k} \leqslant 0$ THEN exit loop.

    3 – Compute the best action α^*;

        IF $\exists \alpha_{e_i} : G(\alpha_{e_i}) > 0$ THEN $\alpha^* \equiv \arg\max_{\alpha_{e_i}} G(\alpha_{e_i})$

        ELSE IF $s_k = (\alpha_{e_1}, \ldots, \alpha_{e_{k+1}}) \wedge V_{s_k} - C_{s_k} > 0$

           THEN $\alpha^* \equiv \alpha_{e_1}$

    4 – Execute α^*, and update the BEL(X_i) vectors;

        Let e such that $\alpha^* = \alpha_e$:

        – execute α^*;

        – $\mathfrak{I} = \mathfrak{I} \cup \{X_e = x_k^e\}$ where k follows from α^*;

        – update BEL(X_i) $\forall X_i$ by propagating inference.

END_LOOP.

7.4 MODELLING SPACE STRUCTURES

Objects and space–time relations are combined into algebraic structures, that are the static components of a visual knowledge base. They act also as *visual memories*, allowing the efficient indexing of stored information.

7.4.1 Modelling objects and relations

Many real-world objects can be represented by recursively decomposing entities into parts and subparts, thus defining *structure*, or *part-of*, *hierarchies*; in addition, the same objects can be viewed as members of abstraction hierarchies (also called *is-a*). Such kinds of entity-relationship representations are naturally formalized in terms of networks (Ballard and Brown, 1982; Roberto, 1994), and Bayes nets provide additional means to account for uncertainty (Binford *et al.*, 1989; Agosta, 1990; Levitt *et al.*, 1990). Two examples of part-of nets are shown in Figure 7.3.

Relations among elements in a scene and an image are to be adequately represented in visual domains (Sucar and Gillies, 1994). To this aim, one can label a network node with a relation name R, with no associated probability distribution (*deterministic link*). A hybrid Bayes net arises, as depicted in Figure 7.3, in which deterministic links are bold. In particular, Figure 7.3(a) represents a part-of net in which the node R encodes the n-ary relation $R(S_1, S_2, \ldots, S_n)$. A probabilistic link connects the object O with R. The matrix of conditional probabilities $M(R|O)$ associated to the latter link represents the probabilities that, given the presence of O in a scene, the relation R holds among the subparts of O. If we now assume that R can be uniquely determined by $(S_1, S_2, \ldots, S_n)$, the net in Figure 7.3(a) is equivalent to the one in 7.3(b).

7.4.2 Time sequences

A vision system operates in a dynamic environment, so that time relations are to be considered, for example, between frame pairs in a sequence. Let us address one such problem, namely the 3D reconstruction of the object O through successive 2D views. Figure 7.4 reports a part-of net in which the occurrence of O in a scene depends on the presence of its subparts (S_i) in successive images; subparts may appear/disappear in different frames.

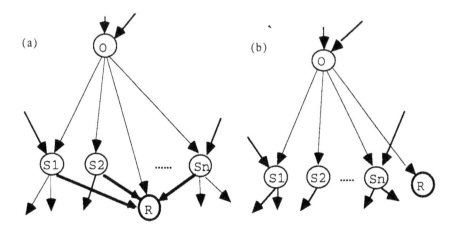

Figure 7.3

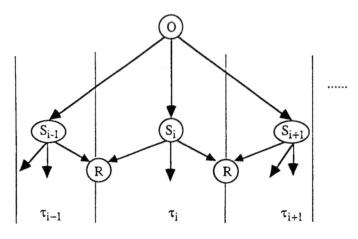

Figure 7.4

7.4.3 Representing actions and tasks

It is interesting to observe that a Bayes net, such as the one in Figure 7.3, can be given a different interpretation, in terms of actions to be performed in order to achieve a goal. This sort of operational semantics of a net greatly enhances its expressive power in representing visual knowledge, and is useful in both static and dynamic problems.

Let us look more closely at Figure 7.3, and suppose that our problem is classifying the object O, i.e., assigning it m possible labels.

An m-variate task variable X_O is assigned to the node O, and is to be instantiated with real values. To do this, we better accomplish simpler tasks: observing at least one of the object subparts/features $S_1, \ldots, S_n$ and/or the relation R. In this way the structural representation of an object in a part-of net becomes a recursive decomposition of the final task into seemingly simpler actions in a new *task net*, until primitive tasks – and the corresponding primitive parts – are put forward. The semantics of probabilistic links is enriched accordingly. The arc labels now quantify how relevant, how promising is performing the subtask T_i with respect to the other ones to achieve the goal, within the context represented by the net. More complex measures of relevance – i.e., based on robustness and computational costs – can be designed for planning purposes, and have been briefly reviewed in Section 7.3.2.

More generally, true *action labels* can be given to nodes, originating new Bayes nets. In the previous example, in order to estimate the feature S_i one may have at disposal, say, m different feature-extraction algorithms. In an action net, each of the latter is associated to a l-valued casual variable, perhaps directly linked to the feature node S_i.

In this case, each link is labelled with a matrix M_{ik}

$$M_k \equiv \begin{pmatrix} p_{11}^k & p_{12}^k & \cdots & p_{1l}^k \\ p_{21}^k & & & \\ \vdots & & & \\ p_{l1}^k & & \cdots & p_{ll}^k \end{pmatrix} \qquad k = 1, 2, \ldots, m$$

where the index i has been omitted everywhere for the sake of clarity. The meaning of the matrix element $\{p_{jq}^{ik}\}$ is: probability that, given the right value x_j^i, the estimate of the kth algorithm is x_q^i. The latter quantifies how precise that estimate is expected to be in all cases. In this way the user may encode his/her knowledge or experience about the performances of processing algorithms, and the latter piece of evidence will be propagated and fused by the net in the current context.

7.5 VISUAL INFERENCE

This section investigates the dynamics of vision in the probabilistic network approach. It is primarily the dynamics of *integration of information* from several sources of evidences and beliefs.

7.5.1 Generalized inference as data fusion

A primary goal of visual processes is *inferring shape and spatial relations* of world entities from sensory evidences. In an artificial system, such a *generalized inference* (i.e., not merely statistical, nor logical) results from integrating (fusing) simpler steps, each of which associating cues and partial models and, also, updating and combining local uncertainty estimates into weighted output hypotheses. The flow of computation is not pre-defined, nor can it be given a definite direction, but top-down and bottom-up partial steps concur, and *perceptual cycles* show up in what has been called (Engelmore and Morgan, 1988) *opportunistic* control scheme.

A sketch of a typical situation – occurring in object recognition, for example – has been shown in Figure 7.5. A Bayes net and its inference mechanism can naturally represent such an integration scheme. The net is divided into several layers. The lowest ones typically refer to elementary, local features extracted from an image; looking at the topology of the net (see also eqn (7.1)), the features are assumed to be independent; they can be instantiated by image processing algorithms. The intermediate nodes represent extended parts (regions) in the image or in a model: at this stage the net performs the matching among evidences and *a priori* knowledge. The highest level nodes correspond to

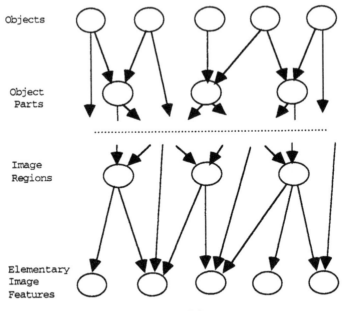

Figure 7.5

the objects that can be located in a scene. The arcs, all oriented in the top-down direction, typically represent *causal relations*, as observed in Section 7.2. It is interesting to observe that probabilistic nets capture the meaning of causality, which is twofold (Binford *et al.*, 1989): on one hand, the presence of an object 'causes' the presence of its projection onto the image plane (*object-percept correspondence*); on the other hand, the object as a whole 'causes' the presence of its parts.

The same perceptual organization in human visual processes, as outlined by the experimental psychologists, can be simulated and represented by probabilistic networks in a comprehensive way; automated systems have been realized accordingly (Sarkar and Boyer, 1993).

7.5.2 Recognition as inference

Let us investigate the role of a net in a very simple form of recognition, which can be viewed as an inference (associative) mechanism. In our example, the (sub)task of the system is detecting the presence of a table in a static scene. There is a part-of net labelled with binary variables, indicating the presence/absence of objects.

The user is expected to provide the *a priori* conditional probabilities (link matrix elements). For example (Figure 7.6), the first column of the matrix describes the possibility that, in the specified environment, legs may be found

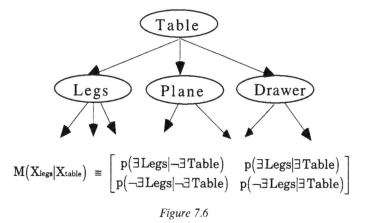

$$M\left(X_{\text{legs}}|X_{\text{table}}\right) \equiv \begin{bmatrix} p(\exists \text{Legs}|\neg \exists \text{Table}) & p(\exists \text{Legs}|\exists \text{Table}) \\ p(\neg \exists \text{Legs}|\neg \exists \text{Table}) & p(\neg \exists \text{Legs}|\exists \text{Table}) \end{bmatrix}$$

Figure 7.6

that do not belong to a table (e.g., there are chairs, or dinosaurs around …).
On the other hand, if in our world we expect that all tables have legs, the
values in the second column will be set near to 1 and 0 respectively. In these
ways, consistently with the basic formalism, the net *accounts for contextual
knowledge*.

Let us set the matrix elements as follows:

$$M(X_{\text{legs}} \mid X_{\text{table}}) = \begin{pmatrix} 0.995 & 0.200 \\ 0.005 & 0.800 \end{pmatrix}$$

$$M(X_{\text{plane}} \mid X_{\text{table}}) = \begin{pmatrix} 0.900 & 0.400 \\ 0.100 & 0.600 \end{pmatrix}$$

$$M(X_{\text{drawer}} \mid X_{\text{table}}) = \begin{pmatrix} 0.700 & 0.400 \\ 0.300 & 0.600 \end{pmatrix}.$$

We now initialize the net by quantifying our 'maximal uncertainty':

$$\text{Bel}(\text{Table}) = (0.5, 0.5).$$

By applying the basic inference algorithm, the initial values are propagated, and
the initial belief values result:

$$\text{Bel}(\text{Legs}) = (0.597, 0.402)$$

$$\text{Bel}(\text{Plane}) = (0.650, 0.350)$$

$$\text{Bel}(\text{Drawer}) = (0.550, 0.450).$$

Suppose that we now observe the legs (i.e., the belief values at the corresponding
node are set accordingly): by propagating the latter values we get a new set of

belief vectors:

$$Bel(Legs) = (1.000, 0.000)$$

$$Bel(Plane) = (0.816, 0.183)$$

$$Bel(Drawer) = (0.649, 0.350)$$

$$Bel(Table) = (0.636, 0.363)$$

and our confidence about the presence of a table is reinforced, as expected.

7.5.3   Inference on shapes and actions

A detailed example illustrates the complex role of inference whenever both objects and processing actions are involved.

Suppose that the task of an inspection system is detecting a feature F in an industrial object: the feature consists of a planar surface with a circular hole of diameter d on top of it (Figure 7.7).

A 2D acquisition system provides images with no perspective distortion (e.g., by performing an orthogonal projection onto the image plane), and can accomplish five visual actions. A *depth-from-stereo* and a *depth-from-focus* algorithm yield evidences about the presence of planar regions in space (set of points at the same depth); a *template matching* module detects a 2D rectangular shape; two distinct modules (*hole detectors* 1 and 2) detect circular holes. We represent the task by the net in Figure 7.8.

The meaning of the casual variables associated to the nodes is:

$X_1 = 0 \rightarrow$ feature F NOT present;   TASK (GOAL) VARIABLE
$X_1 = 1 \rightarrow$ feature F present

$X_2 = 0 \rightarrow$ NO planar surface
$X_2 = 1 \rightarrow$ planar surface of depth z_1
$X_2 = 2 \rightarrow$ planar surface of depth z_2
$X_2 = 3 \rightarrow$ planar surface of depth z_3

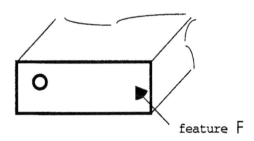

feature F

Figure 7.7

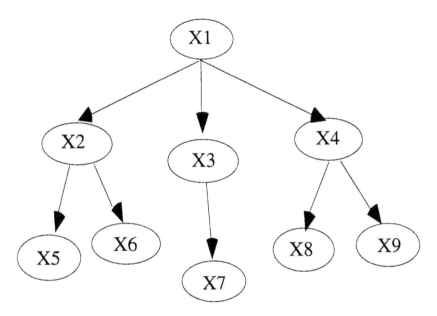

Figure 7.8

$X_3 = 0 \rightarrow$ NO 2D rectangular shape
$X_3 = 1 \rightarrow$ 2D rectangular shape

$X_4 = 0 \rightarrow$ NO circular hole
$X_4 = 1 \rightarrow$ circular hole with diameter $<d - \varepsilon$
$X_4 = 2 \rightarrow$ circular hole with diameter $\geq d - \varepsilon$ and $\leq d + \varepsilon$
$X_4 = 3 \rightarrow$ circular hole with diameter $>d + \varepsilon$

$X_5 = 0 \rightarrow$ action depth from focus detected NO planar surfaces
$X_5 = 1 \rightarrow$ the action detected a planar surface with depth z_1
$X_5 = 2 \rightarrow$ the action detected a planar surface with depth z_2
$X_5 = 3 \rightarrow$ the action detected a planar surface with depth z_3

$X_7 = 0 \rightarrow$ action template matching detected NO shapes
$X_7 = 1 \rightarrow$ the action detected the shape

$X_8 = 0 \rightarrow$ action hole detector 1 did NOT detect a circular hole
$X_8 = 1 \rightarrow$ action detected a circular hole with diameter $<d - \varepsilon$
$X_8 = 2 \rightarrow$ action detected a circular hole with diameter $\geq d - \varepsilon$ and $\leq d + \varepsilon$
$X_8 = 3 \rightarrow$ action detected a circular hole with diameter $>d + \varepsilon$.

The values of X_6 and X_9 are associated to the results of the actions depth from stereo and hole detector 2 respectively, in the same way as X_5 and X_8. The goal variable is X_1, while X_2, X_3, X_4 label subtasks relevant for the solution, and are supposed to be conditionally independent; the variables from X_5 to X_9 are the five actions the system can afford.

We report the connection matrices we used:

$$M(X_2|X_1) = \begin{pmatrix} 0.7 & 0.1 \\ 0.1 & 0.15 \\ 0.1 & 0.6 \\ 0.1 & 0.15 \end{pmatrix} ; \qquad M(X_3|X_1) = \begin{pmatrix} 0.7 & 0.2 \\ 0.3 & 0.8 \end{pmatrix}$$

$$M(X_4|X_1) = \begin{pmatrix} 0.4 & 0.1 \\ 0.2 & 0.15 \\ 0.2 & 0.6 \\ 0.2 & 0.15 \end{pmatrix}$$

$$M(X_5|X_2) = \begin{pmatrix} 0.997 & 0.001 & 0.001 & 0.001 \\ 0.001 & 0.8 & 0.149 & 0.199 \\ 0.001 & 0.15 & 0.701 & 0.2 \\ 0.001 & 0.049 & 0.149 & 0.6 \end{pmatrix}$$

$$M(X_6|X_2) = \begin{pmatrix} 0.997 & 0.001 & 0.001 & 0.001 \\ 0.001 & 0.55 & 0.074 & 0.049 \\ 0.001 & 0.3 & 0.851 & 0.049 \\ 0.001 & 0.149 & 0.074 & 0.901 \end{pmatrix}$$

$$M(X_7|X_3) = \begin{pmatrix} 0.8 & 0.2 \\ 0.2 & 0.8 \end{pmatrix} ; \qquad M(X_8|X_4) = \begin{pmatrix} 0.997 & 0.001 & 0.001 & 0.001 \\ 0.001 & 0.601 & 0.199 & 0.199 \\ 0.001 & 0.199 & 0.601 & 0.199 \\ 0.001 & 0.199 & 0.199 & 0.601 \end{pmatrix}$$

$$M(X_9|X_4) = \begin{pmatrix} 0.997 & 0.001 & 0.001 & 0.001 \\ 0.001 & 0.601 & 0.074 & 0.199 \\ 0.001 & 0.199 & 0.851 & 0.199 \\ 0.001 & 0.199 & 0.074 & 0.601 \end{pmatrix}.$$

A few comments will clarify our choices. The matrices $M(X_2|X_1)$, $M(X_3|X_1)$, $M(X_4|X_1)$ encode the *a priori* knowledge, i.e., how much the presence of F is related to that of its (sub)features. For example, $M(X_4|X_1)$ expresses, in the first column, that, even though F is not present, there may be holes belonging to other parts of the object. In the same way (second column), given that F is actually present, the hole may not be there, or may not have the right diameter (i.e., there may be defects in the object).

The matrices of the arcs connecting intermediate- with low-level nodes in the net encode how reliable are the associated visual actions. For example, observing $M(X_5|X_2)$, $M(X_6|X_2)$ the actions associated to X_5, X_6 are reliable when checking the absence of a planar surface (first column). Moreover, the decreasing diagonal elements in the former matrix express the known property of depth from focus to be more reliable when estimating nearby surface depths; the opposite is true for depth from stereo (increasing diagonal elements).

Let us now explore the rôle of inference in this net by discussing a few cases. We start with setting the belief values of the goal node to maximal uncertainty, and obtain the first set of belief values on the nodes.

- *Case 1* A surface has been observed in the scene at a constant depth z_2, so we set $\text{Bel}(X_2) = (0, 0, 1, 0)$. After propagating the latter values, we find that the belief value corresponding to 'F is present' now is considerably higher, as expected:

$$\text{Bel}(X_1) = (0.143, 0.857)$$

- *Case 2* We now instantiate the variable X_4 (instead of X_2) with the value 2. The belief values of the goal variable – updated by propagating the latter piece of evidence – become:

$$\text{Bel}(X_1) = (0.250, 0.750)$$

The event 'detection of a hole with the correct diameter' has a minor impact on the final goal than the presence of a planar surface.

- *Case 3* A planar surface at depth z_2 – but no holes! – has been detected. The belief values now are:

$$\text{Bel}(X_1) = (0.400, 0.600)$$

The result expresses an even weaker – but still apparent – belief in the presence of the feature F: we interpret the result by saying that, although the object is likely to have a defect, F can still be there.

- *Case 4* We test and compare the impact of the hole detectors 1 and 2. In two separate runs, we calculate the belief values of the goal node after gathering the same evidence – a hole with diameter within the tolerance values – from the first and second detector respectively. The final values are:

$$\text{Bel}(X_1) = (0.323, 0.677)$$
$$\text{Bel}(X_1) = (0.305, 0.695).$$

The system is more confident in the presence of F when applying the second hole detector; this is consistent with the *a priori* knowledge about the performances of the detectors, encoded in the matrices

$$M(X_5|X_2), \qquad M(X_6|X_2).$$

The cases reviewed so far illustrate an interesting property of Bayes nets: the same inference mechanism provides a *basic estimate of relevance of visual actions* which can be further exploited for planning (see the next section).

7.6 PLANNING VISUAL CONTROL

The present section deals with the dynamics of vision in the active approach, in which the main issues concern the control of visual actions.

7.6.1 Selective vision – measures of relevance

In active vision (Aloimonos *et al.*, 1987) a system is expected to modify dynamically its sensing and motion parameters, on the basis of evidences gathered from the environment, and with the aim of accomplishing a task efficiently with respect to the available resources. Therefore, an active approach is a *selective* one, and become of primary importance the issues of planning action sequences, as well as focusing the attention on useful pieces of information. Both issues rely on estimates of relevance, and can be addressed in the framework of Bayes nets and decision theory (Sections 7.3.1 and 7.3.2).

The example in Section 5.3 will better illustrate the latter points.

Let us compute the most promising visual actions according to the EVSI and Shannon measures (Section 7.3.2). To compute the EVSI measure we use eqn (7.4) with the following payoff matrix elements:

$$U = \begin{pmatrix} +1000 & -1000 \\ -2000 & +1000 \end{pmatrix}.$$

Starting with maximal uncertainty the measures of impact of possible visual actions are reported here below:

EVSI measure	Shannon measure
$V(X_1, X_5) = 446.7$	$I(X_1, X_5) = -0.211$
$V(X_1, X_6) = 446.7$	$I(X_1, X_6) = -0.215$
$V(X_1, X_7) = 110.0$	$I(X_1, X_7) = -0.046$
$V(X_1, X_8) = 120.1$	$I(X_1, X_8) = -0.069$
$V(X_1, X_9) = 195.1$	$I(X_1, X_9) = -0.080.$

The two measures agree in indicating the action 'detect planar surface' as the most promising one – which is consistent with the *a priori* information encoded in the matrices $M(X_2|X_1)$, $M(X_3|X_1)$, $M(X_4|X_1)$.

However, the Shannon measure – not the EVSI one! – in this case allows us to discriminate between the two visual actions that can gather evidence on the presence of planar surfaces: the most promising one is *depth from stereo* (X_5). This is because the latter is more reliable in detecting planar surfaces at depth z_2.

Let us now suppose that depth from stereo has detected a planar surface at z_1, $(\text{BEL}(X_6) = (0, 1, 0, 0))$, and analyse the indications from the two measures, provided that three alternative evidences have been collected by hole detector 2.

We report our results and omit the numerical details for brevity:

- *Case 1* The action hole detector 2 has detected no hole (so, $\mathrm{BEL}(X_9) = (1, 0, 0, 0)$). In this case the two measures suggest different actions: according to the Shannon measure, the most promising one is template matching: it is the only action on which no evidence has been collected yet. On the other hand, the EVSI measure suggests to apply hole detector 1 to validate the result provided by hole detector 2, since the latter is not reliable in 'detecting no holes'.

- *Case 2* Hole detector 2 has detected a hole with diameter $< d - \varepsilon$ (so, $\mathrm{BEL}(X_9) = (0, 1, 0, 0)$). The two measures agree in suggesting template matching.

- *Case 3* Hole detector 2 has detected a hole with diameter $\geqslant d - \varepsilon$ and $\leqslant d + \varepsilon$ ($\mathrm{BEL}(X_9) = (0, 0, 1, 0)$). The Shannon measure still indicates the same action. The EVSI measure suggests depth from focus, in order to verify the result yielded by depth from stereo, which is unreliable for small depth values.

The indications provided by the Shannon measure do not vary significantly in the three cases. EVSI indicates as most useful these actions that validate results eventually obtained by unreliable actions.

Finally, it should be observed the Shannon measure is a logarithmic one, which makes it more difficult to interpret and to normalize, especially when used in connection with linear measures (Pearl, 1988; Rimey and Brown, 1994).

7.6.2 The visual planning system TEA-1

A recent, successful application of probabilistic networks to visual problems is the TEA-1 system, by the research team in Rochester (Rimey, 1993; Rimey and Brown, 1994). A detailed account on the whole work is reported in Rimey (1993); we shall briefly consider a few specific issues, relevant to the planning and focus-of-attention purposes.

TEA-1 is a system based on a platform of moving cameras, that solves problems for a large class of scenes (called the *T world*). It adopts a representation scheme in which both domain and control knowledge are made explicit by means of probabilistic networks, complemented with decision-theoretic criteria for rating and selecting relevant information.

Knowledge is represented by four distinct Bayes nets: PART_OF, EXPECTED AREA, IS_A and TASK nets. The former three gather evidences from sensors, and transmit the resulting belief values to the latter – the TASK net – that is the true support to the control steps. There are *visual* and *sensor-positioning* actions.

The control scheme in TEA-1 is based on the iterative selection of the evidence-gathering action that maximizes the expected utility. Therefore, the control cycle is actually that of a planner, and has been designed according to the criteria outlined in Section 7.3. We now summarize its main logical steps:

1. List executable actions.
2. Select the action with highest expected utility.
3. Execute the action.
4. Attach the resulting evidence to the Bayes nets, and propagate its influence.
5. Repeat, until the task is solved.

It can be readily seen that the latter scheme belongs to the class of nonmyopic planners, and a more detailed control algorithm is close to the one reported in Section 7.3.3.

The EXPECTED AREA net represents the areas in the scene in which an object is expected to be located, by means of 2D discrete 32×32 casual variables as node labels. In addition, the geometric relations among objects are encoded as link labels, in the form of conditioned probability matrices $M_{E\_A}(X_j | X_i)$, which mean: probabilities that the object X_j is centered in (x', y'), given that X_i is centered in (x, y) of the visual field.

The role of the net is essential in focusing the attention of the system, since at run-time the size of the expected area of an object is likely to decrease, as more and more evidence is collected on nearby objects.

TEA-1 has been tested extensively, with both simulated and real images, and the results are promising for future developments in visual planning.

7.7 CONCLUSIONS

Probabilistic networks provide a sound and flexible formalism to represent visual knowledge in uncertain – real! – environments. As far as seeing is providing plausible hypotheses about the world, by integrating (fusing) several sources of information with *a priori* models, Bayes networks are good candidates to formalize the underlying complex processes. At the moment, there are no formal schemes offering the same flexibility and robustness.

We summarize the pros and cons of the formalism in the context of visual problems.

Bayes nets meet the adequacy requirements of a knowledge representation model (Reiter and Mackworth, 1989/90), since they provide consistent and complete descriptions in the sense specified in Section 7.2.2.

At the same time, the intractability of pure probabilistic approaches is avoided by fully exploiting the information about correlation among causal variables: in this way, the (global) problem of describing a multivariate probability distribution is factorized into a set of local problems, easier to manage, and restricted to truly correlated variables. The same network topology reflects the properties – correlation, conditional independence – of the variables.

Inference is a key point in Bayes nets: it provides a non-monotonic, distributed, robust mechanism of belief revision. At the same time, inference is a way to propagate and integrate information from several sources in a consistent way. Intelligent functionalities, such as learning and planning, can be realized on the support offered by the network and inference schemes.

Such properties are particularly relevant to visual problems. Bayesian data fusion is now a well-established approach to model uncertainty in multi-sensor acquisition (Clark and Youille, 1990) and low-level vision (Szeliski, 1989) so that Bayes nets provide a natural framework to unify low-level and high-level processing in a comprehensive and sound scheme.

In addition, a network supports a knowledge representation that accounts for space–time structures, useful to represent spatial objects and relations, as well as to allow the efficient indexing of information in visual knowledge bases and memories.

On the other hand, the flow of inference in a net is adequate to formalize the opportunistic flow of control in a typical perceptual problem. Actions can be readily added to the scheme – such as sensor positioning, or feature extraction from input data. Moreover, the net still contains sufficient redundancy to take into account contextual information, and to constrain the flow of visual information accordingly.

We can say that the rich semantics of Bayes nets captures in an affective way the properties of real systems acting in real environments.

However, probabilistic nets still suffer from considerable limitations. Inferencing on a generic net is a NP-complete problem (Cooper, 1990), so that constraints must be imposed on the topology (e.g., nets must be trees or multi-trees). Even in trees, it is hard to handle the amount of information required, e.g., when dealing with multiple-valued variables (Agosta, 1990; Rimey and Brown, 1994): as a matter of fact, the formalism is viable only for a small number of casual variables.

An additional limitation is that in most nets the designer is expected to provide the *a priori* probability values necessary to initialize the net: the automated learning mechanisms are still subject of research (Sucar and Gillies, 1994). Finally, the network topology relies upon a number of assumptions of independence which may be hard to verify, especially in visual domains.

However, the results of the past and current research work are encouraging. We believe that the progress in both theoretical and experimental work on Bayes nets for vision will yield remarkable results towards the solution of more and more complex perceptual tasks.

REFERENCES

Agosta, J.M. (1990) The structure of Bayes networks for visual recognition. In: *Uncertainty in artificial intelligence 4* (eds R.D. Shachter, T.S. Levitt, L.N. Kanal and J.F. Lemmer), 397–405. Elsevier, Amsterdam, The Netherlands.

Aloimonos, Y., Weiss, I. and Bandyopadhyay, A. (1987) Active vision. *Proc. of the First IEEE Conf. on Computer Vision*, 35–54.

Ballard, D. and Brown, C.M. (1982) *Computer vision*. Prentice Hall, Englewood Cliffs, New Jersey.

Bhatnagar, R. and Kanal, L.N. (1993) Structural and probabilistic knowledge for abductive reasoning. *IEEE Trans. on Pattern Analysis and Machine Intelligence* **PAMI-15**, 3, 233–245.

Binford, T.O. (1982) Survey of model-based image analysis systems. *International Journal of Robotics Research* **1**, 1, 18–64.

Binford, T.O., Levitt, T.S. and Mann, W.B. (1989) Bayesian inference in model-based machine vision. In: *Uncertainty in artificial intelligence 3* (eds L.N. Kanal, T.S. Levitt and J.F. Lemmer), 73–95. Elsevier, Amsterdam, The Netherlands.

Breese, J.S. (1992) Construction of belief and decision networks. *Computational Intelligence* **8**, 4, 624–647.

Brooks, R.A. (1983) Model-based three-dimensional interpretation of two-dimensional images. *IEEE Trans. on Pattern Analysis and Machine Intelligence* **PAMI-5**, 2, 140–150.

Charniak, E. (1991) Bayesian networks without tears. *AI Magazine* **12**, 4, 50–63.

Clark, J.J. and Youille, A.L. (1990) *Data fusion for sensory information processing systems*. Kluwer, Boston, Massachusetts.

Cooper, G.F. (1990) The computational complexity of probabilistic inference using bayesian belief networks. *Artificial Intelligence* **42**, 393–405.

Dagum, P. and Chavez, R.M. (1993) Approximating probabilistic inference in Bayesian belief networks. *IEEE Transactions on Pattern Analysis and Machine Intelligence* **PAMI-15**, 3, 246–255.

Dean, T.L. and Wellman, M.P. (1991) *Planning and control*. Morgan Kaufman, Los Altos, California.

Engelmore, R.S. and Morgan, A.J. (eds) (1988) *Blackboard systems*, Addison Wesley, Wokingham, UK.

Goldman, R.P. and Charniak, E. (1993) A language for construction of belief networks. *IEEE Trans. on Pattern Analysis and Machine Intelligence* **PAMI-15**, 3, 196–208.

Heckerman, D., Horvitz, E. and Middleton, B. (1993) An approximate nonmyopic computation for value of information. *IEEE Trans. on Pattern Analysis and Machine Intelligence* **PAMI-15**, 3, 292–298.

Henrion, M. (1988) Propagating uncertainty in Bayesian networks by probabilistic logic sampling. In: *Uncertainty in artificial intelligence 2*, 149–163. Elsevier, Amsterdam, The Netherlands.

Jensen, F.V., Christensen, H.I. and Nielsen, J. (1992) Bayesian methods for interpretation and control in multi-agent vision systems. In: *Applications of artificial intelligence X: machine vision and robotics*. SPIE Proceedings Series, Vol. 1708.

Levitt, T.S., Agosta, J.M. and Binford, T.O. (1990) Model-based influence diagrams for machine vision. In: *Uncertainty in artificial intelligence 5* (eds M. Henrion, R.D. Shachter, L.N. Kanal and J.F. Lemmer), 371–388. Elsevier, Amsterdam, The Netherlands.

Neapolitan, R.E. (1990) *Probabilistic reasoning in expert systems*. Wiley, New York.

Pearl, J. (1986) Fusion, propagation, and structuring in Bayesian Networks. *Artificial Intelligence* **29**, 3, 241–287.

Pearl, J. (1987a) Evidential reasoning using stochastic simulation of causal models. *Artificial Intelligence* **32**, 245–257.

Pearl, J. (1987b) Addendum: evidential reasoning using stochastic simulation of causal models. *Artificial Intelligence* **33**, 131.

Pearl, J. (1988) *Probabilistic reasoning in intelligent systems: networks of plausible inference.* Morgan Kaufmann, San Mateo, California.

Peot, M.A. and Shachter, R.D. (1993) Fusion and propagation with multiple observations in belief networks. *Artificial Intelligence* **48**, 3, 299–318.

Reiter, R. and Mackworth, A.K. (1989/90) A logical framework for depiction and image interpretation. *Artificial Intelligence* **41**, 125–155.

Rimey, R.D. (1993) Control of selective perception using Bayes nets and decision theory. Technical Report 468, Department of Computer Science, University of Rochester.

Rimey, R.D. and Brown, C.M. (1994) Control of selective perception using Bayes nets and decision theory. *International Journal of Computer Vision* **12**, 2/3, 173–207.

Roberto, V. (1994) Models and descriptions in machine vision. In: *Human and machine vision, analogies and divergences*, (ed. V. Cantoni), 245–274. Plenum Press, New York.

Sarkar, S. and Boyer, K.L. (1993) Integration, inference, and management of spatial information using Bayesian networks: perceptual organization. *IEEE Trans. on Pattern Analysis and Machine Intelligence* **PAMI-15**, 3, 256–274.

Schachter, R.D. and Peot, M.A. (1990) Simulation approaches to general probabilistic inference on belief networks. In: *Uncertainty in artificial intelligence 5*, (eds M. Henrion, R.D. Shachter, L.N. Kanal and J.F. Lemmer), 221–231, Elsevier, Amsterdam, The Netherlands.

Sucar, L.E. and Gillies, D.F. (1994) Probabilistic reasoning in high-level vision. *Image and Vision Computing* **12**, 1, 42–60.

Szeliski, R. (1989) *Bayesian modeling of uncertainty in low-level vision*, Kluwer, Boston, Massachusetts.

Verri, A. (1996) The regularization of early vision. *This volume.*

8

Distributed Systems for Fusion of Visual Information

8.1 DISTRIBUTED SYSTEMS AND INFORMATION FUSION

A distributed system (DS) is composed of a collection of *passive* and/or *active* elements, connected by functional relations, supporting the transmission and analysis of distributed information.

Example of passive DSs, that have in common a model of propagation of the information in a medium, are acoustic pipes and electric lines. In case of telegraph lines, the Maxwell equations hold as a theoretical model, and the corresponding distributed elements are: inductance ($l\,\mathrm{d}x$) and capacitance ($c\,\mathrm{d}x$) per unit length, as shown in Figure 8.1.

By taking the limit $\mathrm{d}x \to 0$ and using the existing relations between c, l, v (shunt voltage) and i (series current), the following wave equations describe the behavior of the distributed system:

$$\frac{\partial v}{\partial x} = -l\,\frac{\partial i}{\partial t} \qquad \frac{\partial i}{\partial x} = -c\,\frac{\partial v}{\partial t}.$$

ARTIFICIAL VISION
ISBN 0-12-444816-X

Copyright © 1997 Academic Press Ltd
All rights of reproduction in any form reserved

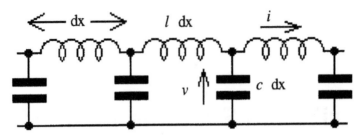

Figure 8.1   An example of distributed system: the telegraph line.

The existence of active (nonlinear) elements allows to design active transmission lines, whose functions recall those of the axons in the nervous system. In the present case, the DS performs distributed computation on its elements. For example, adding a negative shunt admittance $-g\,dx$ to the line in Figure 8.1 determines real solutions of the corresponding wave equation of the form:

$$v \text{ or } i \sim \mathrm{Re}\,[e^{\pm\gamma x + j\omega t}] \text{ where } \gamma = \left(-\frac{g}{2}\sqrt{\frac{l}{c}} + j\omega\sqrt{lc}\right)$$

where $\mathrm{Re}(\,)$ stands for the real part of the wave equation; the $\pm$ sign at the exponent represents waves propagating in the $\pm x$ directions, respectively. In the example, the information is transmitted and transformed by the DS. The latter kinds of active networks have been studied in Scott and Johnson (1969) and Scott (1970), to design the first examples of active retinas. They have been realized by means of Josephson junction transmission lines, that consist of two superconductor strips separated by an insulating layer, which permits the coupling of the superconducting wave functions or tunnelling of superconducting electrons.

The previous examples of DSs are composed by continuous elements. Nevertheless, distributed computation can be performed also on discrete distributed systems. Given a network of simple computing elements, the most straightforward scheme is to use one computing element (CE) for each *action*, which is called *local computation*. The information description and its representation can be also distributed in the network at a local level. Relations among the elements can be again represented at the element level. The evolution of a discrete DS can be described by a difference equation system, or by a set of transition functions. The latter formalism is adopted in automata systems and Petri nets.

Visual data are spatially distributed; complementary informations are related to spatial measurements (visual model, sensors description), and the analysis is carried out combining heterogeneous sources of information: DSs are natural candidates to perform heterogeneous analysis. A mesh of processors is a simple example of DS dedicated to image analysis (see Figure 8.2) (Duff, 1976; Batcher, 1980; Miller and Stout, 1985). The CEs are locally connected to match the

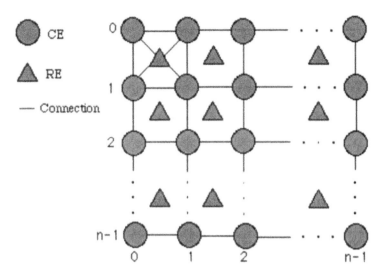

Figure 8.2 An example of mesh-connected DS.

topological distribution of data. The computational complexity of each CE is tuned to satisfy its functional purposes. In the example in Figure 8.2, circles represent computing elements, while triangles are relational elements (RE) devoted to communication purposes. An RE may contain information about the relations between the CEs. Both CEs and REs are arranged in a 2D array, and addressable by a pair of indexes (i, j).

The evolution of a mesh-connected DS can be formulated in the framework of finite difference wave equations, the information flow depending on the initial conditions and their constraints.

For example, the transition rules describing the wave propagation in Figure 8.3, can be stated as follows:

$$\Delta_x CE_{i,j} = CE_{i-1,j} - CE_{i,j}$$

$$\Delta_y CE_{i,j} = CE_{i,j-1} - CE_{i,j}$$

$$CE_{i,j} = \Delta_x CE_{i,j} + \Delta_y CE_{i,j}$$

$$CE_{i-1,j} = 0$$

$$CE_{i,j-1} = 0$$

for $T_{k+1} = T_k + 1$ and $1 < k < 8$

Here the time is discrete. The initial condition is $CE_{0,0} = 1$ at $T_1 = 0$. The constraint confines the computation inside the mesh array. Such an approach has been used to model the heat propagation equation. This kind of physical process is at the basis of the solution of several relaxation labeling algorithms. Again, the use of active CEs allows the design of active vision systems and smart retinas (Tallen *et al.*, 1988; Zavidovique and Bernard, 1990; Brown, 1992).

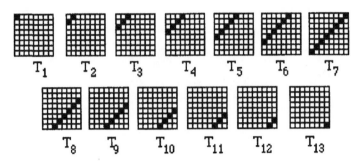

Figure 8.3   Wave propagation in a mesh topology.

Information fusion   The model of distributed computation is suitable to handle algorithms that integrate more than one source of information (Gerianotis and Chau, 1990; Pinz and Barth, 1992). In fact, each element (either active or passive) may contain fragment of information to be transmitted or analysed. In the literature this topic is usually addressed as data fusion. In the following, the term information fusion will be used to enlarge the concept of fusion to: data, algorithms, knowledge and models (Campbell *et al.*, 1990; Campbell and Cromp, 1990; Flachs *et al.*, 1988). This wider definition includes all ingredients used in computer vision.

In vision problems, several acquisition and processing methods lead to a large amount of multiple data sets. For example, a scene can be considered with different lighting, multiview can be provided by a multi-sensors system (Gerianotis and Chau, 1990; Hedengren *et al.*, 1990; Duncan *et al.*, 1987; Hanson *et al.*, 1988) and many applications are based on multi-spectral data (Di Gesù, 1994; Di Gesù and Romeo, 1994). The multi-resolution approach also provides a set of data at several levels of definition: from pixel measurements to region properties. Another example of multi-data is the acquisition of multitemporal data of the same subject taken at different times (Hong and Brzakovic, 1990; Hong, 1992; Mason, 1992; Chao *et al.*, 1990).

Usually, each component of the multiple data set contributes unique information; however, the whole information is affected by a high degree of redundancy among the components. A fusion of these components is, under many aspects, profitable if the following goals are reached:

- - a reduction of the amount of data;
- the results are more robust;
- scene description is more complete.

Information fusion techniques are becoming relevant in computer vision tasks (Shulman and Aloimonos, 1990; Tahani and Keller, 1990; Shen, 1991; Krishnapuram and Lee, 1992; Keller *et al.*, 1992). For example, Burt (1988) proposes an approach to system control called 'dynamic vision' based on the integration of information. It is inspired by the visual attention mechanisms in humans, and consists of three elements: foveation, tracking and high-level interpretation. This attention mechanism supports efficient analysis by focusing the system's sensing and computing resources on selected areas of the scene. Moreover, the resources are rapidly redirected by the evolution of the scene and tasks themselves.

Having one single image accomplishing multiple tasks (or one task with changing parameters) results in a new set of images, or descriptors of them: such data sets can also be fused. Meer *et al.* (1990) investigates in depth how outputs of multiple, slightly-varying image processing tasks can be cumulated; the process is named 'consensus vision'. The consensus image is more robust than any single-output image.

8.2 FUSION OF VISUAL INFORMATION

Owing to the limitations of a single sensor and the uncertainties of sensor data, fusing information from a multi-sensor system is used to acquire a representation of the environment. The issues addressed are: how to represent the information; how to fuse it; and how to control fusion.

8.2.1 Fusion of multi-sensor data

We start by examining the sequence of processes involved in a typical computer vision system as an information fusion mechanism (Pinz and Bartl, 1992). The scene is a small portion of the 'world', i.e., it is a narrow section of a high-dimensional (3d space + time) world. By exposure, we obtain a 2D image; digital image processing techniques and pattern recognition methods are applied on it to extract features that form an image description. An image understanding system generates a scene description. Finally, several scene descriptions are integrated into a world description. This linear sequence of processes (from *scene selection* to *scene understanding*) becomes a cycle when dealing with robots, for example (Figure 8.4).

The previous example allows us to introduce several *levels* of description (world level, scene level, . . .), and at each level – except the world level – several instances of sources exist. The sources of information are called: *external*, when they are generated by sensors (multiple scene and multiple images); *internal*, if they are produced by a multi-tasks procedure or by a single task with different parameter values.

In Figure 8.5, the tree is shown of the multiple sources of information related to computer vision problems. The information grows from the *world* apex to the *world description*, because of both multiple tasks results and multiple scene views.

In Figure 8.6 the fusion mechanism is described as a combination of the

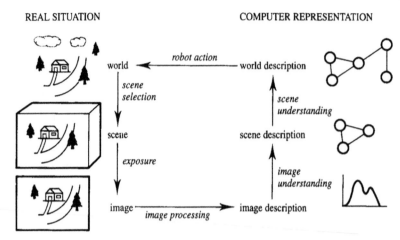

Figure 8.4   Sequence of processes in computer vision.

previous sources. Furthermore, it is possible that new components are generated during the fusion process. Both scene and image level are considered
external, because their acquisition is given by sensors, while the description
levels are internal, because they are due to the internal computation. This
distinction is not really a strict one; in fact, an active vision system that uses
smart retinas, performs processes at the acquisition level.

8.2.2   Integrated decision techniques

Integrated decision techniques can be described by using graph-theoretical
approaches. The graphs are weighted and directed; the nodes are labelled and
represent:

- algorithms used in the decision phase (A);
- input data (D);

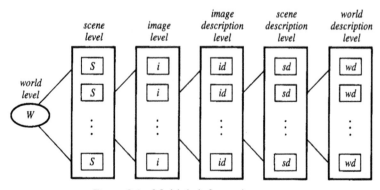

Figure 8.5   Multiple information sources.

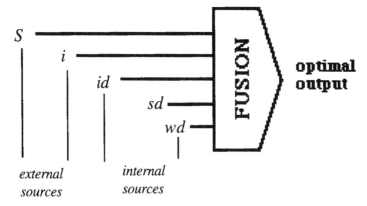

Figure 8.6   The process of fusion.

- output data (O);
- models (M).

Three strategies to optimize decision making are here described: *serial*, *parallel* and *network*. The decision algorithms may be based on min–max recursive rules, and heuristics. In addition, Bayesian and Dempster–Shafer inference procedures have been considered to perform decisions.

Let $A \equiv \{A_1, A_2, \ldots, A_s, \ldots, A_S\}$ be a set of algorithms, and Π_A an initial possibility distribution defined on A:

$$\Pi_A = (A_1/\pi_1, A_2/\pi_2, \ldots, A_s/\pi_s, \ldots, A_S/\pi_S) \qquad 0 < \pi_i \leqslant 1$$

The symbol A_i/π_i denotes the assignment of a possibility value π_i to the algorithm A_i. Note that $\Sigma_i \pi_i$ could be greater than 1. The meaning of Π_A depends on the experimental situation; for example, it could be related to the accuracy of each algorithm, or could be determined by the user on the basis of his/her experience. In the latter case, it can be interpreted as a belief or a probability distribution, depending on the nature of the uncertainty: subjective or not. The output of each A_i is indicated by O_i.

An Integrated Fuzzy Algorithm (IFA) consists in evaluating the final output, O, and its related possibility value, π, by means of functions denoted $G(O, D, M; \Pi_A)$, and $m(\Pi_A)$ respectively. The algorithms may interact and interchange information in several fashions, here three main strategies are considered: the *serial combination* (IFA-serial), the *parallel combination* (IFA-parallel) and the *network combination* (IFA-net).

IFA-serial   The algorithms are logically arranged in a chain. Each A_s, with $1 < s \leqslant S$, receives three inputs: (O_{s-1}, D, M_s), and produces the result O_S. In the case of chain combination (Figure 8.7), the output, O_s, and the related possibility value π_s, at the sth stage depend on both the algorithm A_s and the result obtained by the $(s-1)$th stage. The evolution of an IFA-serial is

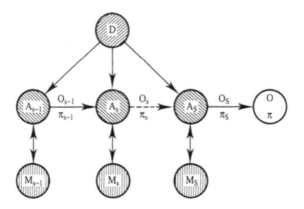

Figure 8.7   ISF-serial integration.

determined by the following system:

$$O_s = G(O_s, O_{s-1}, M_s, \pi_s, \pi_{s-1}) \qquad \pi_s = m(\pi_s, \pi_{s-1}) \qquad \text{for } 1 < s \leqslant S$$

$$O = O_S \qquad\qquad\qquad\qquad \pi = \pi_S.$$

Note that in an IFA-serial procedure the updating of π_s is performed after the computation of O_s. It must be pointed out that the result may depend on the order in which the As are executed. Moreover, serial combination must be preferred whenever the algorithms interact sequentially to find, step by step, the solution.

Examples of functions G and m are:

$$O_s = (O_s \pi_s + O_{s-1}\pi_{s-1})/(\pi_s + \pi_{s-1})$$

$$\pi_s = \max\{(\pi_s + \pi_{s-1})/2, \pi_{s-1}\} \qquad \text{for } 1 < s \leqslant S.$$

IFA-parallel   The algorithms perform their computations independently (see Figure 8.8). Each A_s, with $1 < s \leqslant S$, receives two inputs: (D, M_s), and produces the result O_s. The evolution of a IFA-parallel is determined by the following system:

$$O = G(O_1, O_2, \ldots, O_S; \pi_1, \pi_2, \ldots, \pi_S) \qquad \text{for } 1 < s \leqslant S$$

$$\pi = m(\pi_1, \pi_2, \ldots, \pi_S) \qquad\qquad \text{for } 1 < s \leqslant S.$$

The functions G and m are usually homogeneous polynomials:

$$O = \sum_{s=1}^{S} \pi_s^{r_s} O_s \qquad \pi = \sum_{s=1}^{S} \alpha_s \pi_s^{r_s} \qquad \text{with} \qquad \sum_{s=1}^{S} \alpha_s = 1.$$

where the exponent r_s is a positive real number, that allows modelling of the influence of each A_s. The final result is then a weighted sum of the partial ones.

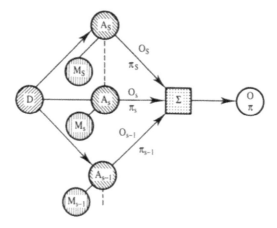

Figure 8.8 ISF-parallel integration.

Examples of functions G and m are:

$$O = \frac{\sum\limits_{s=1}^{S} \pi_s O_s}{\sum\limits_{s=1}^{S} \pi_s} \qquad \pi = \frac{1}{S} \sum\limits_{s=1}^{S} \pi_s.$$

The parameter π is usually considered as a global measure of performance of the IFA-parallel procedure. The IFA-parallel approach should be preferred whenever the algorithms do not exchange information during the computation.

IFA-net This is the most general case of algorithm interconnection. The graph representing the network must satisfy three conditions:

- only one source node, labelled D, exists;
- only one sink node, labelled O, exists;
- at least one direct path between D and O exists.

Note that hybrid combination strategies can be designed; for example, an IFS algorithm could be represented by a connected graph, in which parallel and serial combination are sub-graphs (see Figure 8.9).

The functions G and m are computed using the serial and parallel rules, depending on the local topology of the network.

IFA algorithms may include backtracking modules in order to optimize the evaluation of Π_A (Pinz and Bartl, 1992). In this case, the computation of the

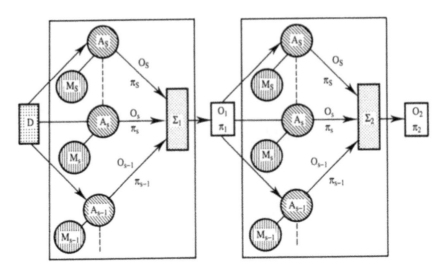

Figure 8.9   An example of IFA-net (parallel/serial integration).

functions G and m is iterated until a cost function $F(\pi_1, \pi_2, \ldots, \pi_S)$ reaches a minimum.

The minimization of the function F can be treated as a combinatorial optimization problem (Garfinkel, 1985), for which several classical methods have been proposed. For example, local search algorithms (Lin, 1965) are based on a stepwise improvement of the value of the cost function by exploring the neighborhood. The Metropolis algorithm can be also considered, its design being borrowed from solid state physics (Barker and Henderson, 1976), to obtain low energy states of a solid in a heat bath. The latter process, named *simulated annealing*, is developed in two steps:

- increase the temperature of the heat bath to a maximum value at which the solid melts;
- decrease carefully the temperature of the heat bath, until the particles arrange themselves in the ground state of the solid.

The thermal equilibrium is characterized by the Boltzmann distribution, which gives the probability of the solid being in a given state s with energy E at temperature T, and is proportional to: $\exp\{-E/kT\}$. In our case, F plays the role of E, and π_s corresponds to the temperature, T.

The cost function can be expressed as follows:

$$F(\pi_1, \pi_2, \ldots, \pi_s) = \frac{1}{2}\sum_{s=1}^{S}\pi_s^2.$$

The Metropolis algorithm is outlined below. It accepts in input an initial evaluation of Π_A and F as computed by the IFA procedure. The array PF indicates the possibility function, variables D and E indicate positive constants. A complete review on simulated annealing algorithms can be found in (Aarts 1990).

```
main( )
  {
  repeat
  {
  IFA;
  Metropolis( )
    {
    F0 = F;
    for (s = 1; s ⇐ S; s++)
      if (exp{−F} > random)
        {
        PF[s] = PF[s] + D;
        recompute(F);
        }
      else
        {
        PF[s] = PF[s] − D;
        recompute(F);
        }
    }
  until (|F0 − F| < E);
  }
```

Bayesian decision The uncertainty inherent to data and methods can be handled including Bayesian rules in the integration phase (Cheesman, 1983). Bayesian rules hold if we assume known the *a priori* probabilities of the output $\{\pi_1, \pi_2, \ldots, \pi_s, \ldots, \pi_S\}$, and the conditional probabilities of the integrated output O, supposing true the output O_i, $\pi_{i/0}$, for $1 \leqslant i \leqslant S$ (Duda and Hart, 1976).

For IFA-parallel, the decision rule can be formulated as follows:

$$O \equiv O_i \Leftrightarrow \pi_{i/0} = \max_{1 \leqslant j \leqslant S} \left\{ \frac{\pi_{0/j} \pi_j}{\pi_0} \right\} \text{ where } \pi_0 = \sum_{k=1}^{S} \pi_{0/k} \pi_k.$$

In case of ignorance, the conditional probabilities $\pi_{i/0}$ are assumed all equal to $1/S$, and the previous rule becomes simply:

$$O \equiv O_i \Leftrightarrow \pi_{i/0} = \max_{1 \leqslant j \leqslant S} \{\pi_j\}.$$

For IFA-serial, the decision rule can be formulated as follows:

$$O \equiv O_i \qquad \text{for } i = 1$$

$$O \equiv O_i \Leftrightarrow \pi_{i/i-1} = \max\left\{\pi_i, \frac{\pi_{i-1/i}\pi_i}{\pi_{i-1}}\right\} \qquad \text{for } 1 < i \leqslant S.$$

In case of ignorance, the conditional probabilities $\pi_{i-1/i}$ are assumed all equal to 1/2, and the previous rule becomes simply:

$$O \equiv O_i \qquad \text{for } i = 1$$

$$O \equiv O_i \Leftrightarrow \pi_{i/i-1} = \max\left\{\pi_i, \frac{\pi_i}{2\pi_{i-1}}\right\} \qquad \text{for } 1 < i \leqslant S.$$

Dempster–Shafer decision   The Bayesian approach assumes the 'a priori' knowledge of probability models, in such a way that it is possible to build exact models of phenomena starting from experimental data, and then use the models to make predictions.

Human experience may play a fundamental role whenever data are plagued by vagueness and methods are not known. Moreover, *cloudy* quantities exist which are not describable in terms of probability distributions, such as the 'beauty of something' or the 'tallness of a person'. In all such cases education, fashion and global knowledge play a crucial role for making decisions.

The Dempster–Shafer rule can be applied (Shafer, 1976). Suppose that two pieces of information π_i and π_j are assigned (e.g., in a subjective way).

Their combination can be expressed by the orthogonal sum:

$$\pi_i(A) \oplus \pi_j(A) = \frac{\displaystyle\sum_{r,s} \pi_{i/r}\pi_{j/s}}{1 - \displaystyle\sum_{h,k} \pi_{i/h}\pi_{j/k}}$$

where $\pi_{i/r}$ and $\pi_{j/s}$ are the conditional probabilities related to information non-contradictory with A, while $\pi_{i/h}$ and $\pi_{j/k}$ are the conditional probabilities related to the contradictory information.

Heuristic decision   In practice, several other empirical functions can be used to integrate partial results; heuristics does not ensure an optimal solution. An example of empirical integration will be shown in the section dedicated to applications.

Computational note   Integrated algorithms are computationally expensive, because they require the execution of S procedures. However, the advent of parallel machines, based on the VLSI technology, provides new tools in order to design faster parallel clustering algorithms. IFA are suitable for parallelization on MIMD (Multi-Instructions Multi-Data) machines.

IFA-serial can be implemented by using a pipeline architecture, with S

processing units (PUs). In this case the sth stage of the pipe executes the algorithm A_s and receives the pairs (A_{s-1}, π_{s-1}) from the previous stage. The last element of the pipe contain the final grouping.

IFA-parallel can be implemented on a multi-processor machine with S PUs, on which the $A_1, A_2, \ldots, A_S$ are executed asynchronously. Moreover, each PU can be a multiprocessor machine with an appropriate network topology, which depends on the algorithm to be executed. The integration phase is performed as soon as all partial results are available.

In Di Gesù (1992) parallel algorithms for image segmentation are reviewed, the computational performances on MIMD multi-processors machines is discussed for both parallel and serial integration techniques.

8.3 DISTRIBUTED INFORMATION SYSTEMS

An extensive literature has grown during the last decade on the problem of distributed architecture for sensor fusion, with application to robotics and remote sensing (Clement *et al.*, 1992; Poloni, 1993; Di Gesù *et al.*, 1993). All the architectures share the following features: the computation is distributed among specialized processing units; knowledge, data and their uncertainties are represented in a distributed fashion (Sinha and Neu, 1989).

We start with an example of a multi-specialist architecture as proposed in Poloni (1993); it has the multi-layered organization shown in Figure 8.10, and has been applied to remote sensing purposes. The example has been chosen since it includes most of the basic components of a distributed cooperating information system.

The design is based on the blackboard concept. The blackboard is used as a shared base of facts, and is accessed via a uniform protocol. Two types of specialists have access to the blackboard:

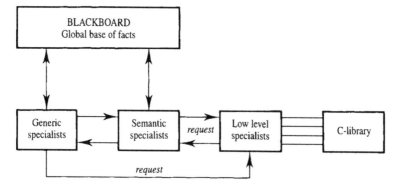

Figure 8.10 The multi-specialist architecture.

- The generic specialists – application independent – perform general scene analysis (e.g. object detection and relation).
- The semantic object specialists – application dependent – are dedicated to high level vision (make hypotheses on an object). Requests are sent to specialized subtasks.

At the third level of specialists feature extraction (skeletonization, region extraction, filtering) is performed; the image processing modules are triggered upon request from the first two specialists, while the low-level specialists, on their side, can invoke the execution of algorithms in a C library.

The system kernel is based on SMECI generator of expert systems (ILOG and INRIA, 1991), and the NMS multispecialist shell (Corby *et al.*, 1990).

8.3.1 Architecture of a distributed system

Definition 1   A DS is a five-tuple $\langle S, I, O, \Delta, W \rangle$, where S is the set of states, $I \subset S$ the set of inputs, $O \subset S$ the set of outputs. Transitions among states are represented by the function $\Delta: P(S) \to P(S)$, where $P(S)$ is a partition of S. W is a mask function $W: S \times S \to \{0, 1\}$.

Visual computation, based on information fusion, can be formulated in terms of five functional modules: *observe, process, world model, choose next* and *action*. Sensor data are provided to the observe module that performs early vision tasks; the information flows to the process module, processes of which (algorithms executed on proper hardware) are selected by the choose-next module, the processes are also driven by the world model. The outputs are directed both to the action, that operates on the environment, and to the observe module, that drives further sensor-explorations. The world represents the environment on which a DS operates. Within the system, information flows in a continuous and active feedback loop (Figure 8.11).

The definition of DS can be interpreted by considering as states:

$$S \equiv D \cup M \cup A \cup P$$

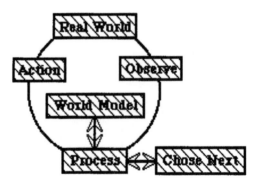

Figure 8.11   A vision system based on information fusion.

where: D represents input/out *data*, collected by sensors from the *world* or produced as results by a given computation; M represents *models* (e.g., relations between objects, sensors, environment); P is the set of distributed processes, and A the set of actions that modify the world (open/close a door, activate an alert system, ...). Moreover, $I \subseteq D$ is the set of input states; $O \subseteq D \cup A$ is the set of output states. Here, the term process has a wide meaning: it could be a single algorithm or a sequence of processes associated to a set of processing units (PU) connected in a reconfigurable topology; in this sense the process is time-dependent.

The transition function has been defined as follows:

$$\Delta = \begin{cases} \Delta_1 : \mathcal{P}(D) \to \mathcal{P}(P) & \bullet \ \text{data/processes} \\ \Delta_2 : \mathcal{P}(P) \to \mathcal{P}(D) & \bullet \ \text{processes/data} \\ \Delta_3 : \mathcal{P}(M) \to \mathcal{P}(P) & \bullet \ \text{models/processes} \\ \Delta_4 : \mathcal{P}(P) \to \mathcal{P}(M) & \bullet \ \text{processes/models} \\ \Delta_5 : \mathcal{P}(P) \to \mathcal{P}(P) & \bullet \ \text{processes/processes.} \end{cases}$$

Each function, Δ_i, defines logical (or physical) links among elements of S, and P is the set of parts of S. The computation evolves on the basis of both transitions and mask functions. A transition is active iff all of its input links have mask value equal to 1. When a transition $P(X) \to P(Y)$ is active, information (data, model and tasks) flows from X to Y, the nature of the information flow depending on the sets X, Y. For example: if $X \in \mathcal{P}(D)$ and $Y \in \mathcal{P}(P)$ data are sent from X to Y; if $X, Y \in \mathcal{P}(P)$ both data and tasks can be sent from X to Y.

In Figure 8.12 a DS is sketched. Note that the arcs are labelled, and the information flow is enabled when the labels w_i are 'on'.

The evolution of a distributed process on a DS can be driven by firing rules. In the next section, the solution adopted for the M-VIF machine (Machine Vision based on Information Fusion) (Di Gesù *et al.*, 1993) will be described.

8.3.2 The M-VIF machine

The M-VIF machine is an example of DS oriented to vision problems. Its architecture is based on a *Compound Node* (CN), which is made of 4 functional modules (Figure 8.13):

- the C module is the controller of the CN, responsible for the evolution of the computation;
- the modules H are dedicated to data processing;
- the modules IP manage the input/output of data;
- the module LN (Link Network) is dedicated to the interconnection of the CNs, in order to realize reconfigurable network topologies.

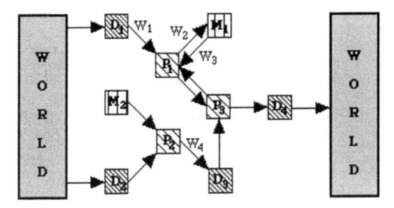

Figure 8.12   Automata for a vision system.

The emulation of M-VIF has been carried out on a reconfigurable and heterogeneous architecture based on the HERMIA machine. It includes 16 general purposes PUs (INMOS T800) and a bank of 6 digital signal processors (INMOS A110).

The system operates in a pipeline modality. For our purposes, we use only one compound node, where H_1 and H_2 have four processing units and are dedicated to the segmentation and matching phases, respectively. H_3 performs the integration and decision phases and requires only one processing unit.

The image I/O is handled by the controller, that loads the data in the shared memory; a bench of memory data units directs data and intermediate results to the appropriate processes.

The evolution of a distributed process is based on *firing* conditions, that must be verified at the input of each state $x \in S$. For this purpose, two sets are

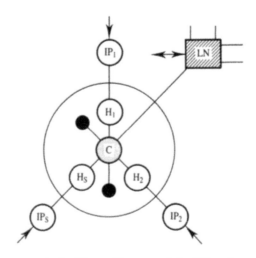

Figure 8.13   The architecture of the CN node.

introduced for each $x \in S$ to define firing rules:

$$IN(x) = \{(s, x) \,|\, s \in S\} \qquad OUT(x) = \{(x, s) \,|\, s \in S\}.$$

Each input is partitioned according to:

$$IN(x) = I_1(x) \cup I_2(x) \cup I_3(x) \cdots \cup I_m(x) \text{ and } I_i(x) \cap I_j(x) = \emptyset \text{ for } i \neq j.$$

Each element of $I_i(x)$ is determined in a unique way by an integer index i_j for $j = 1, 2, \ldots, k_i$. To each $I_i(x)$ a logic function $f_i(x)$ is associated:

$$f_i(x) = \left(\bigwedge_{j,1}^{k_i} w(i_j, x) \right) \wedge \overline{\left(\bigvee_{r \neq i} \left(\bigwedge_{j,1}^{k_r} w(r_j, x) \right) \right)}$$

Here, $w(i_j, x)$ is the mask value of the transition (i_j, x). From the previous definition follows that $f_i(x)$ is 1 if and only if each transition $j \in I_i(x)$ is '1' and each $f_r(x)$ with $r \neq i$ is '0'. The *firing* rule can be now stated as follows: *fire* x iff $\exists I_i(x)$ such that $f_i(x) = 1$.

The introduction of the mask function f_i and the partition induced by $IN(x)$ and $OUT(x)$, allow to implement the $\{\Delta_i\}$ functions holding in the DS.

The computation performed by M-VIF depends on the values of the mask function, w_0, at the starting time, t_0, i.e. for each x, the values $w(i_j, x)$ are set initially to '0' and '1'.

The updating of the mask values is determined by those elements of S that, firing at a given time, assign new mask values to the elements of $OUT(x)$. Therefore, the value w_{i+1} at time t_{i+1}, depends on both the values w_i, and the results at time t_i. The computation ends as soon as the condition of *firing* is false for all the elements of S. Note that the evolution of the whole mask values is determined by the deterministic tasks running in each state x.

8.3.3 Visual distributed environment

Humans interact with the real environment through their senses, and react and make decisions depending on the result of such interactions. This observation has suggested most of the interactive computer systems based on virtual reality and multi-media. Vision plays a relevant role among the human senses, and efforts have been done in the last decade to improve the design of visual interfaces for computer systems.

The performance of a visual system relies upon the ability to focus the areas of interest, by maximizing a given costs/benefits utility criterion (Swain and Stricker, 1991; Brown, 1992). The *selection* of interesting regions is relevant, and the ability to select *salient* features is the basic question in artificial and natural intelligence. Moreover, visual perceptual systems should be able to adapt their *behavior* depending on the current goal and the nature of the input data. Such performances can be obtained in systems able to interact dynamically with the environment.

DSs are characterized by a huge amount of states and parameters, distributed through several elementary sub-system units. Moreover, complex functional dependencies exist between states and parameters. Automated control assessment and motion detection in risky environments are examples of distributed systems. In these cases, visual data are usually collected from several sensors, and elaboration is carried out on local processing units, logically interconnected to share and interchange knowledge (models, data and algorithms).

A Visual Distributed Environment (VDE) must provide a synthetic view of a DS behavior, and guides to understand local and/or global computation phases. In fact, the design and implementation of algorithms on multi-processor machines depends on the distribution of data and processes among the units; then, the dynamic control is relevant to optimize and tune the execution of processes. Dynamic visual tools allow to realize such user/machine interaction in a natural and efficient way.

The VDE design can be based on a multi-layered graph grammar (Chang and Kunii, 1981; Pfeiffer, 1990). In the following, the VDE implementation will be shortly described on the first emulated version of the machine M-VIF. The concept of dynamic icon (Di Gesù and Tegolo, 1992) will be used to extend visual interfaces on DSs. Informally, a VDE is an icon-based system that allows to build a user view of the underlying DS, to allocate and control the related cooperating processes. Bricks of a VDE are both conventional and Dynamic Icons (DI). The semantic values of a DI depend on the evolution of the distributed processes. In order to define more formally the VDEs, it is necessary to introduce some notations and definitions about distributed systems.

Definition 2   A dynamic icon $\gamma \in DI$ is a correspondence between a set of metaphors, M, and a set of icons, I; in the present context, metaphors have a perceptual meaning, since they may represent visual patterns as well as acoustic signals or a combination of both.

The formal definition of VDE can now be stated as follows:

a) $M_m: M \to P$ (metaphor – process)
b) $M_p: P \times M \leftrightarrow DI$ ((process, metaphor) – icon)
c) $M_\Delta: DI \times DI \to \Delta$ ((icon, icon) – transition functions)
d) $M_w: DI \times DI \to \{0, 1\}$ (mask flag on the relational arcs).

M_m is a function that assigns a metaphor to a process, is a many-to-one correspondence, since the corresponding process reaches different configurations during the evolution of the DS; M_p is one-to-one, in order to avoid ambiguities – it is responsible for the assignment of a γ. The pair (P_i, m_k) determines the current value of the icon $\gamma_i^{(k)}$; M_Δ defines the relations between γs, and has been introduced to handle the evolution of a DS at a visual level; the function M_w assigns a mask flag value to each relational arc. The mask flag is useful to visualize the information flow in the DS. Figure 8.14 shows the diagram of the functions introduced above. The previous definition is useful to represent the evolution of a process in a DS.

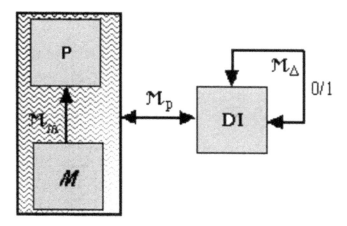

Figure 8.14 Relations between the sets DI, *P*, and *M*.

A process p_i defined as a sequence of virtual processes:

$$(p_i^{(k)}|k = 1, 2, \ldots, N)$$

corresponding to a sequence of dynamic icons

$$(\gamma_i^{(k)}|k = 1, 2, \ldots, N),$$

where:

$$\gamma_i^{(k)} = M_p(p_i^{(k)}, m^{(k)}) \qquad \text{for } k = 1, 2, \ldots, N.$$

Visually, a VDE can be represented by direct graphs G, the nodes of which are the γs, and the labelled arcs are determined by the function M_Δ; the label values λ are determined by the kind of Δ function ($\lambda = 1, 2, 3, 4, 5$), and the mask value m is fixed by M_m. Moreover, a dynamic icon γ may represent recursively a sub-set of the VDEs, named VDE$_\gamma$. Such kind of dynamic icon is named a *compound*. This feature makes it easier to design and program a DS.

Two compounds, γ_1 and γ_2, are linked by a *compound arc* $(M_\Delta(\gamma_1, \gamma_2))$ iff a subset of arcs, with equal labels, exists between the corresponding sub-graphs VDE$_{\gamma_1}$ and VDE$_{\gamma_2}$. The mask flag of a compound arc is set on the basis of AND-OR rules applied to the mask-flags of the corresponding arcs. Dynamic icons and arcs, which are not compound, are said to be *primitive*.

The introduction of compound and primitive dynamic icons makes it possible to organize a VDE in a hierarchical way (Figure 8.15), and this approach is useful to handle a DS at different levels of refinement. The hierarchical structure of the VDE allows to develop and update the visual design of a DS, and to focus the attention easily where errors and bugs occur. Moreover, a VDE controls the evolution of a DS at different levels of detail (from compound to primitive).

The evolution and the programming of a DS can be exploited via a VDE by testing the syntactic correctness of visual graphs. Visual parsing is recursively applied to compound elements, until primitive elements are reached. The parsing phase is accomplished by graph grammars.

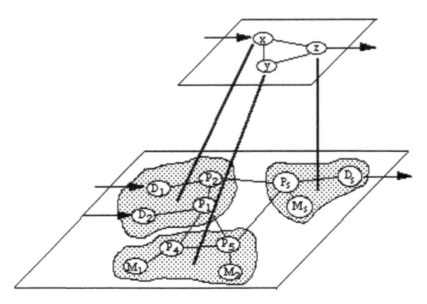

Figure 8.15 Conceptual organization of a VDE.

A graph representing a VDE must satisfy the following properties, that are used during the parsing phase:

a)   the graph G of a VDE is connected;
b) G has one input and one output labelled arcs;
c)   valid sub-graphs of G have input/output arcs labelled with the same label;
d)   direct paths exist from input to output nodes of G and its valid sub-graphs.

The automatic inspection of a VDE, the proper direction of an arc can be easily tested. Moreover, during the visual editing some semantic inconsistencies can be discovered; for example, it is not allowed to input or output processes, data and models that do not match correct prototypes.

The whole syntactic correctness of a DS is tested during the parsing phase of the VDE. The parsing consists of match and merge procedures applied to the graph G, that can be considered as a visual program. The *match* step tests the consistency of subgraphs G' of G; for this purpose, standard graph-matching algorithms can be used; their computational complexity is reduced to $O(L)$, where L is the number of arcs, because of the constraints imposed by the graph grammar. The *merge* step creates a super-node, the input (output) arcs of which are the input (output) arcs of G'. The parsing is successful if a single super-node is obtained, at the end of this phase; the merge step depends on the number of nodes in G. The consistency rule depends on the graph grammar, and how it has been defined when constructing the visual program.

In the following, the VDS design of the M-VIF machine is outlined. It is based on three sets of visual elements: icons, metabase (metaphor database), and arcs.

• *Icons* Dynamic icons representing processes are defined by the pair (color, pattern), and share the same visual organization (Figure 8.16). The body contains a visual pattern and an expansion button, the color indicating also the current status (active, wait, stopped) of the icon. The primitive icons have a yellow background, the compound icons a blue one. Six input/output channels are foreseen to implement the *IN/OUT* lines of each node. The expansion of a compound icon is performed by pushing its button. The expansion of a primitive icon depends on the kind of connected information. For example, if an icon represents a process, it returns the source code in I-PICL language (Di Gesù and Tegolo, 1992), while it returns kernel values if it represents a kernel. The distribution of the resources (processors, sensors, data knowledge) are handled by the user. This is obtained by creating a link between the appropriate dynamic icon and the corresponding resource.

• *Metabase* Is a database of metaphors which contains visual, acoustic and text patterns. The dynamic evolution of a process is represented by three colors, default values are:

green, for 'active';
orange, for 'wait';
red, for 'stopped'.

The metabase can be updated by the user.

• *Arcs* Arcs are drawn by colored lines, joining icons, each color being related to a Δ function: yellow (Δ_1), orange (Δ_2), blue (Δ_3), red (Δ_4) and black (Δ_5). Mask values are represented by dotted ($m = 0$) and continuous ($m = 1$) lines. Primitive and compound arcs differ for the thickness of the lines: bold (compound), thin (primitive).

The user interacts with the VDS by using a mouse or touch-screen tools. He/she may develop distributed procedures on M-VIF, at three levels of abstraction:

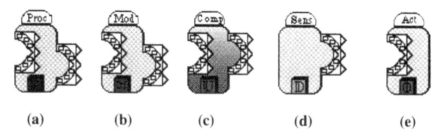

(a)   (b)   (c)   (d)   (e)

Figure 8.16   Dynamic icons: a) process; b) model; c) compound; d) input data; e) output data.

- by combining compound visual elements (dynamic icons) already defined;
- by combining primitive and compound icons;
- by developing I-PICL procedures.

Visual tools are also provided for connecting metaphors to processes and assigning primitive icons to (process, metaphor) pairs. A first prototype of VDS is under development under Windows 3.1 and the C++ object oriented language.

8.4   APPLICATIONS

We end this chapter with some applications of information fusion techniques to image analysis and processing.

8.4.1   An integrated method for image retrieval

One of the main challenges in a pictorial database (Tanaka and Ichikawa, 1988; Zink and Jaffe, 1993) is retrieving an image or a sub-image starting from the user's pictorial description obtained by a sketch or by example from a prototype. Unfortunately, much information is included inside this representation, and most of it doesn't help to understand the meaning of an image via automated inspections. Another approach uses text annotations that outline the image content, and their management leads to the image retrieval. The problem of this approach is determined by the fact that '1000 words aren't enough to describe even a small picture'. This causes the non-applicability of such methods to experimental databases. Retrieval tasks become even harder when multi-dimensional data are involved.

The application of data fusion techniques to image retrieval regards both picture representation and textual information, to focus the context of the retrieval (biomedical, astronomical, archaeological, ...). In Tegolo (1994) a retrieval technique is proposed, based on information fusion. The retrieval is performed in three steps (Figure 8.17):

- *Segmentation*   The input image is segmented by four clustering algorithms: Hierarchical Single Link Clustering (HSLC), Hierarchical Histogram Partition Clustering (HHPC), Hierarchical ISODATA (HISO) and Two Phases Clustering (TPC). In this step, each method selects a set of candidates $(S_1, \ldots, S_4)$ to be compared with a given target image (P) in the matching step.

- *Matching*   Two distance functions (Euclidean, Hausdorff) are used in a cooperating way to reach the best matching solution. The results $S_1, \ldots, S_4$ are analysed by the corresponding matching modules $MM_1, \ldots, MM_4$. Each

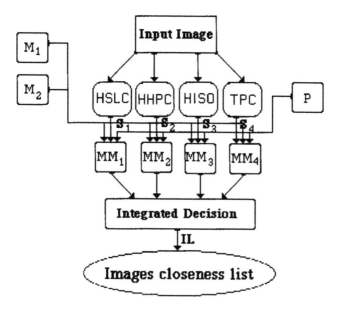

Figure 8.17 General design of the retrieval system.

module MM_i uses three different input data: metrics (M_1, M_2), the segmented image (S_i), and the target image P, and provides an ordered list of the retrieved objects (candidates). The matching is performed on the basis of several shape features: size, skewness, moments. Ordering is performed on an evaluation parameter (φ), ranging in the interval $[0, 1]$, related to each candidate.

- *Integrated decision* It performs the best retrieval by combining the list of candidates LM_1 and LM_2, computed in the matching phase, in the list IL. The integration rule adopted is based on a weighted ranking, expressed by the following weight function:

$$
\alpha_j =
\begin{cases}
-\dfrac{j}{K} + 1 & \text{for } 0 \leqslant j < \left\lceil \dfrac{K}{2} \right\rceil \\[2ex]
\dfrac{j}{K} & \text{for } \left\lceil \dfrac{K}{2} \right\rceil \leqslant j \leqslant K.
\end{cases}
$$

where j is the position in the lists, and K is the cardinality of the lists. Let j_1 and j_2 be the entry points of the segment c in LM_1 and LM_2 respectively, then the entry point, j, in the integrated list IL is:

(a)
$$ j = \alpha_j j_1 + \alpha_j j_2; $$

(b)
$$ j = \left\lceil \frac{j - j_{min}}{j_{max} - j_{min}} K \right\rceil $$

where j_{max} and j_{min} are the minimum and maximum entry points computed following the rule (a). This choice allows to avoid integration techniques involving non-uniform parameters, and gives a predominant weight to segments with highest or lowest ranks inside the lists.

The retrieval method has been applied to an astronomical database, containing sky maps and planets photos. Figure 8.18 shows an example of retrieval. The input image (Figure 8.18a) is a part of the moon's surface; the prototype image (Figure 8.18b) has been extracted from the left bottom corner. Four copies of the prototype, with different size and orientation, have been added to the input image to produce the test image (Figure 8.18c).

8.4.2 Integrated clustering for MRI segmentation

Magnetic Resonance Imaging (MRI) plays a relevant role in the design of systems for computer assisted diagnosis. MR images are multi-spectral data, so that the physician has to combine several sources of perceptual information to perform the tissue classification useful for the final diagnosis. The number and kind of tissue classes depend on the anatomical target. For example, in the case of MRI of the human skull, eight classes are considered relevant: background, bone, cerebrospinal fluid, fat, grey matter, haemorrhage, tumor and white matter. Manual classification allows the experts to discriminate quite easily soft tissues for the diagnosis of diseases inherent to that kind of the cerebral matter (cancer, haemorrhage). Other classes of diseases (neoplastic, ischemic, ...) are less evident and need the integration of visual information coming from all the bands.

Four feature vectors can be associated to each pixel of an MR image: T_1, T_2 (spin-echo image), TI (inversion recovery image), and GRASS (Gradient-Recalled Acquisition in the Steady-State image). Therefore, each pixel x of an image M can be described by six parameters (two spatial coordinates and four spectral values): $x \equiv (i_x, j_x, \eta_1, \eta_2, \eta_3, \eta_4)$.

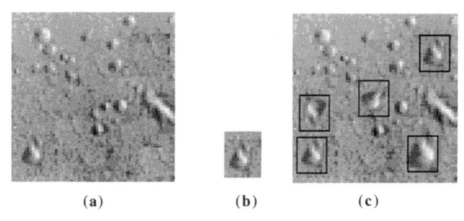

(a)   (b)   (c)

Figure 8.18 (a) The input image; (b) the prototype; (c) the retrieved candidates.

The methodology has been integrated in an image analysis system for the analysis of MR images, that utilizes information of different nature: measurements, knowledge base, algorithms and pictorial database. All these are combined in making decisions.

The automatic clustering system for MRI-image analysis proposed by Di Gesù and Romeo (1994) is based on an information fusion methodology, that allows to perform the final segmentation by combining multi-spectral MRI data and the results of four segmentation algorithms.

Hierarchical Single-Link Clustering (HSLC)   The algorithm is based on a top-down approach. Starting from the input X, a tree structure is computed, the leaf nodes representing the final clustering of the data set, the inner nodes the image segmentation at some level of refinement. Three elements must be taken into account during the computation of the tree clustering: the choice of the split rule, the consistency test in order to stop the splitting of a node, and the labelling of each leaf node.

Hierarchical Histogram Partition Clustering (HHPC)   The algorithm performs the recursive dichotomy of the input image data. During the computation a binary tree is built, the leaves of which represent a possible clustering of X. The elements to be considered during the computation of the tree clustering are: the choice of the split rule, the consistency test, and the labelling of leaf nodes. The probability distribution of the cluster intensities is directly estimated by using the histogram distribution of the intensity levels in the four bands.

Hierarchical ISODATA (HISO)   HISO also performs a recursive dichotomy of the input image. It is based on the application of the k-mean partition algorithm. The input is a set of N patterns, each pattern being a column vector with L components: $X = [x_1, x_2, \ldots, x_i, \ldots, x_N]$ with $x_i \equiv (x_{i,1}, x_{i,2}, \ldots, x_{i,L})$, the barycenter of each cluster, k, is denoted by B_k. X is assumed metric and $\rho(i, k)$ represents the distance between the element x_i and the barycenter, B_k, of the cluster C_k. HISO applies the k-mean algorithm setting $K = 2$ at each node of the classification tree. The algorithm stops as soon as the required number of clusters or segments is reached. The leaves of the tree represent the segments of the input image.

Two-Phases Clustering (TPC)   The algorithm is composed of two steps: (a) link each pixel, x, to the nearest-neighbor, y, iff $\rho(x, y) \leq \phi$, where ρ is a distance function and ϕ is a positive threshold (after this phase the image is segmented in N_ϕ clusters); (b) merge the N_ϕ clusters, by using an intercluster similarity function based on a variant of the Mahalanobis distance.

Integrated Clustering (IC)   IC has been tested on 4-bands Magnetic Resonance Images (8 bit-per-pixel intensity values, and 224×256 pixels per image) detected by a 1.5 Tesla imaging system (GE Sigma) at the Department of

Radiology of the Stanford University Hospital. They represent MRI of the human skull, with the eight relevant classes quoted above. The training set was based on a set of 20 images analysed by a team of three radiologists, who performed the manual classification. Let $M^{(i)}$ be the partition performed by the algorithm i, and $0 \leqslant \pi_i \leqslant 1$ be a numeric value that estimates the validity of the method. In our case, this value has been obtained experimentally from the confusion matrix ($\pi_{\text{TPC}} = 0.63$, $\pi_{\text{HHPC}} = 0.77$, $\pi_{\text{HSLC}} = 0.6$, $\pi_{\text{HISO}} = 0.73$). The IC algorithm has been implemented by using a global strategy. The final classification, λ, has been assigned by mean of the weighted vote function, evaluated in each pixel:

$$\lambda(x) = \max_{1 \leqslant k \leqslant 8} \left\{ \sum_{i=1}^{4} \pi_i \delta(k, \lambda_i(x)) \right\}$$

where $\lambda_i(x)$ is the label class assigned to x by the algorithm 'i', and δ is a proximity function (in our case is the Dirac delta function).

Table 8.1 reports the percentage, P_e, of agreement with the human classification. It is interesting to note that the percentage was $p_e = 95\%$, in the case of the IC-global procedure. This result indicates that the integration of more then one method can produce better results than each method taken separately.

The same data set has been analysed with two methods: AFCM (De La Paz et al., 1987) and the classical ISODATA procedure provided by the ISAW system (Hanson and Myers, 1988). The IC algorithm seems to show a better global accuracy.

8.4.3 Fusion of stereo and motion for 3D reconstruction

This application regards a fusion technique based on stereo and motion information for 3D reconstruction; details may be found in Waldman and Merhav (1992). The approach has been developed to minimize the risk of detection of a rotocraft operating in a high-threat environment, by providing the necessary range estimates for obstacle avoidance and autonomous navigation. The surrounding terrain, vegetation, and man-made obstacles are used in

Table 8.1

Method	P_e
TPC	0.63
HISO	0.73
HSLC	0.6
HHPC	0.77
IC	95
AFCM	88
ISODATA	87

conjunction with passive sensing to assure concealment in the nap of the earth flight. Piloting under these circumstances is a demanding task. Unfamiliar images, such as provided by a forward-looking infra-red camera or low-light level television, often cause misinterpretation of visual cues. The camera motion and position are assumed to be provided by an inertial navigation system aided by a global positioning system. It allows image derotation, improved feature matching and the estimation of absolute depth. In Figure 8.19 the general design of the integrated vision system is shown.

The *motion module* is based on a bank of independent extended Kalman filters, each using depth as the state and the matching of corner features as the measurement. During the recursive depth estimation the dynamic of the measurement noise is considered, by deriving analytically the correlation of sequential measurement noise. However, this procedure does not avoid the computation of the filter covariance, which is an ill-posed problem.

The motion module, in spite of the optimal solution assured by the use of extended Kalman filters, has a limited accuracy due to the region close to the focus of expansion of the camera. In fact, the measured corner shift is too small within this region, and signal/noise matching ratio becomes unfavorable. Furthermore, the accuracy is degraded by the occurrence of the aperture problem.

The *stereo module* allows an instantaneous estimation of the measurement

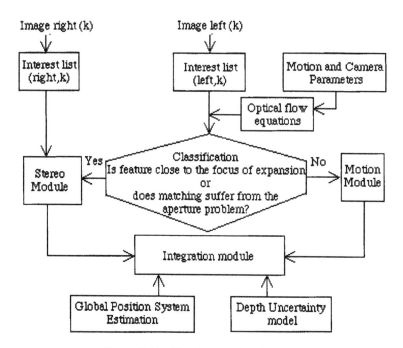

Figure 8.19   The integrated vision system.

noise, by performing stereo matching on alternative search areas. The stereo module relates to the image array as a discrete 2D signal. Thus, the measured disparity has its resolution limited by the image resolution. The conditional probability density of depth given the measured stereo disparity is derived and the minimum variance and the maximum *a posteriori* depth estimators are then computed.

The depth estimates from the motion and stereo modules provide a pair of sparse depth maps which convey information about the surroundings at isolated 3D locations. These estimates are fused within a volumetric representation (Elfes, 1987; Elfes and Matthies, 1987) which assumes that accurate position estimates are available from a global positioning system. Thus, the depth estimates relative to the camera are superimposed on the local north-east-down reference frame. These estimates are regarded as evidence of voxel occupancy.

The *fusion module* provides also an uncertainty model for the depth estimates, which is used to update the measurements of occupancy and emptiness of the voxels. A raytracing algorithm computes the intersections between the voxels and the lines of sight of the estimate 3D locations. A set of rules based on spatial reasoning is devised to discard erroneous evidence, to reinforce the correct ones, and to consider the possibility of previously existing evidence which became occluded due to camera motion. Different combinations of occupancy and emptiness measurements are used to define voxels as obstacles.

8.5 CONCLUSIONS

This chapter was intended to provide a general background on information fusion and distributed systems, oriented to image analysis and computer vision. Intelligent engines (robots, monitoring and alert systems, . . .) collect information from heterogeneous sensors (acoustic, visual, tactile, . . .) and their computation is heterogeneous too. The pursual of an engine goal is centered on combining such information; moreover, the complexity of the algorithms involved needs multi-processor systems, on which heterogeneous computation can be performed efficiently. Several questions are still under investigation. For example, the reconfigurability of distributed systems seems to be one of the requirements for handling information fusion, in order to adapt the system to the evolving environment. However, more effort is to be done to obtain *real time* reconfigurability. The need exists of designing new optimal strategies for scheduling tasks, and of tuning the system to the algorithm requirements. An additional open problem regards the design of general combination strategies in case of heterogeneous information: the current combination rules are not always satisfactory, so that heuristics and *ad hoc* solutions, derived from experience, are often adopted.

REFERENCES

Aarts, E. and Korst, J. (1990) *Simulated annealing and Boltzmann machines*. Wiley, New York.

Barker, J.A. and Henderson, D. (1976) What is liquid? Understanding the states of matter. *Reviews of Modern Physics* **48**, 587–671.

Batcher, K.E. (1980) Design of a massively parallel processor. *IEEE Trans. on Computers* **C-29**, 836–840.

Brown, C.M. (1992) Issue in selective perception. *Proc. 11th IAPR Int. Conf. on Pattern Recognition*, A, 21–30.

Burt, P.J. (1988) Attention mechanism for vision in a dynamic world. *Proc. 9th Int. Conf. on Pattern Recognition* 977–987.

Campbell, W.J. and Cromp, R.F. (1990) Evolution of an intelligent information fusion system. *Photogrammetric Engineering and Remote Sensing* **56**, 6, 867–870.

Campbell, W.J., Cromp, R.F., Hill, S.E., Goettsche, C. and Dorfman, E. (1990) Intelligent information fusion for spatial data management. In *Proceedings of the 4th International Symposium on Spatial Data Handling*, **2**, 567–78, Zurich, Switzerland.

Chang, S. and Kunii, T. (1981) Pictorial data-base systems. *IEEE Computer*, **14**.

Chao, J.J., Cheng, C.M. and Su, C.C. (1990) A moving target detector based on information fusion. *Proc. of the IEEE 1990 International Radar Conference*, 341–344.

Cheesman, P. (1983) A method of computing generalized Bayesian probability values for expert systems. *Proc. 8th Int. Joint Conf. on Artificial Intelligence*. 198–202.

Clement, V., Giraudon, G. and Houzelle, S. (1992) A multi-specialist architecture for sensor fusion in remote sensing. In *Proc. of the 11th CPR*, The Hague, A, 202–206.

Corby, O., Allez, F. and Neveu, B. (1990) A multi-expert system for pavement diagnosis and rehabilitation. *Transportation Research Journal*, **24** A(1).

De La Paz, R.L., Bernstein, R., Hanson, W.A. and Walker, M.G. (1987) Approximate Fuzzy C-Means (AFCM) cluster analysis of medical magnetic resonance image (MRI) data. *IEEE Trans. on Geoscience and Remote Sensing* **GE-25**, 815–824.

Di Gesù, V. (1992) Parallel algorithms for image segmentation. *Proc. of 16th ÖAGM-Meeting*, 184–198, Vienna, A.

Di Gesù, V. (1994) Integrated fuzzy clustering. *Fuzzy Sets and Systems* **FSS 68**, 3, 293–308.

Di Gesù, V. and Romeo, L. (1994) An application of integrated clustering to MRI segmentation. *Pattern Recognition Letters* **15**, 731–738.

Di Gesù, V. and Tegolo, D. (1992) The iconic interface for the pictorial C language. *Proc. of IEEE Workshop on Visual Language* 92, IEEE Comp. Soc. Press, 119–124, Seattle, Washington.

Di Gesù, V., Gerardi, G. and Tegolo, D. (1993) M-VIF: A machine-vision system based on information fusion. *Proc. of CAMP'93* (eds M.A. Bayoumi, L.S. Davis and K.P. Valavanis) 428–435, IEEE Comp. Soc. Press, New Orleans, Louisiana.

Duda, R.O. and Hart, P.E. (1976) *Pattern classification and scene analysis*. Wiley, New York.

Duff, M.J.B. (1976) CLIP4: a large scale integrated circuit array parallel processor. *Proc. 3rd International Joint Conf. Pattern Recognition*.

Duncan, J.S., Gindi, G.R. and Narendra, K.S. (1987) Low-level information fusion: multisensor scene segmentation using learning automata. *Proc. Spatial Reasoning and Multi-Sensor Fusion*, 323–331, Morgan Kaufmann, Los Altos, California.

Elfes, A. (1987) Sonar-based real world mapping and navigation. *IEEE Trans. on Robotics and Automation* **RA-3**, 3, 249–265.

Elfes, A. and Matthies, L. (1987) Sensor integration for robot navigation: combining sonar and stereo range data in a grid-based representation. *Proc. of the 26th Conf. on Decision and Control*, 1802–1807.

Flachs, G.M., Jordan, J.B. and Carlson, J.J. (1988) Information fusion methodology. *Proc. of Sensor Fusion*, SPIE – The International Society for Optical Engineering, **931**, 56–63.

Garfinkel, R.S. (1985) Motivation and modeling. In: *The traveling salesman problem* (eds E.L. Lawler, J.K. Lenstra, A.H.G. Rinnooy Kan and D.B. Shmoys), 17–36. Wiley, New York.

Gerianotis, E. and Chau, Y.A. (1990) Robust data fusion for multisensor detection systems. *IEEE Trans. on Information Theory* **36**, 6, 1265–1279.

Hanson, W.A. and Myers, H.J. (1988) *Image Science and Applications Workstation (ISAW 1.10) user's guide*, IBM Scientific Center, Palo Alto, California.

Hanson, A.R., Riseman, E.M. and Williams, T.D. (1988) Sensor and information fusion from knowledge-based constraints. *Proc. of SPIE*, **931**, 186–196.

Hedengren, K.H., Mitchell, K.W. and Ritscher, D.E. (1990) Information fusion methods for coupled eddy-current sensors. *Proc. of SPIE* **1198**, 334–345.

Hong, L. (1992) Recursive temporal-spatial information fusion. *Proc. of the 31st IEEE Conference on Decision and Control* **4**, 3510–3511.

Hong, L. and Brzakovic, D. (1990) An approach to 3D scene reconstruction from noisy binocular image sequences using information fusion. *Proc. of the Third Int. Conf. on Computer Vision*, 658–661.

Keller, J.M., Krishnapuram, R. and Hobson, G. (1992) Information fusion via fuzzy logic in automatic target recognition. *Proc. of SPIE* **1623**, 203–208.

Krishnapuram, R. and Lee, J. (1992) Fuzzy-set-based hierarchical networks for information fusion in computer vision. *Neural Network* **5**(2), 335–350.

ILOG and INRIA (1991) *SMECI 1.54, Le manuel de référence*, Gentilly, F.

Lin, S. (1965) Computer simulations of the traveling salesman problem. *Bell System Technical Journal* **44**, 2245–2269.

Mason, K.P. (1992) Information fusion in a knowledge-based classification and tracking system. *Proc. 5th Int. Conf. on Industrial and Engineering Applications of AI and Expert Systems*, 666–675. Springer-Verlag, Berlin, Germany.

Meer, P., Mintz, D., Montanvert, A. and Rosenfeld, A. (1990) Consensus vision. *Proc. of the AAAI-90 Workshop on Qualitative Vision*, 111–115.

Miller, R. and Stout, Q.F. (1985) Geometric algorithms for digitized pictures on a mesh-connected computer. *IEEE Trans.* **PAMI-7**, 216–228.

Pfeiffer, J.J. (1990) Using graph grammars for data structure manipulation. In *Proc. of the 1990 IEEE Workshop on Visual Languages*, Skokie, 42–47.

Pinz, A. and Bartl, R. (1992) Information fusion in image understanding. *Proc. 11th Int. Conf. on Pattern Recognition* (ICPR), A 366–370.

Poloni, M. (1993) A fuzzy-based architecture for sensor integration and fusion. *Proc. of EUFIT'93 Congress on Fuzzy Intelligent Technologies*, ELITE Foundation, 188–194.

Scott, A.C. (1970) *Active and non-linear wave propagation in electronics.* Wiley, New York.

Scott, A.C. and Johnson, W.J. (1969) Internal flux motion in large Josephson junctions. *Applied Physics Letters* **14**, 316–318.

Shafer, G. (1976) *A mathematical theory of evidence.* Princeton University Press, Princeton, New Jersey.

Shen, S.S. (1991) 3D-scene interpretation through information fusion. *Proc. of SPIE* **1382**, 427–433.

Shulman, D. and Aloimonos, J. (1990) Probabilistic foundations for information fusion with applications to combining stereo and contour. *Proc. of SPIE* **1198**, 356–360.

Sinha, D. and Neu, D.J. (1989) A general class of aggregation operators with applications to information fusion in distributed systems. *Proc. of the IEEE Int. Conf. on Systems, Man and Cybernetics*, **3**, 921–927.

Swain, M.J. and Stricker, M. (eds) (1991) *Promising directions in active vision.* Technical Report CS 91-27, University of Chicago.

Tahani, H. and Keller, J.M. (1990) Information fusion in computer vision using the fuzzy integral. *IEEE Trans. on Systems, Man and Cybernetics* **20**, 3, 733–741.

Tallen, T., Faggin, F., Gribble, G. and Mead, C.A. (1988) Orientation-selective VLSI retina. *Proc. SPIE Int. Symp. on Visual Communications and Image Processing*, 1040–1046, Cambridge, Massachusetts.

Tanaka, M. and Ichikawa, T. (1988) A visual user interface for map information retrieval based on semantic significant. *IEEE Trans. on Software Engineering* **14**, 5, 666–670.

Tegolo, D. (1994) Shape analysis for image retrieval. In: *SPIE Storage and Retrieval for Image and Video Database II* (eds W. Niblack and R.C. Jain), 59–69. San Jose, California.

Waldmann, J. and Merhav, S. (1992) Fusion of stereo and motion for 3D reconstruction. *Proc. of the 11th Int. Conf. on Pattern Recognition* (ICPR), The Hague, A, 5–8.

Zavidovique, B.Y. and Bernard, T.M. (1990) Smart retinas. *Proc. Cognitiva-90*, 495–515.

Zink, S. and Jaffe, C. (1993) Medical imaging database. *Investigat. Radiology* **28**, 4, 366–372.

9

Hybrid Computation and Reasoning for Artificial Vision

9.1   INTRODUCTION

The design and the realization of intelligent autonomous systems, able to operate in unstructured environments, is one of the most significant goals of applied artificial intelligence and related fields, such as advanced robotics. The ability to interact autonomously with the external world requires an adequate development of the perceptive capabilities and the internal representations of the environment, in such a way to allow the drawing of inferences and the capability of decision making of the system.

Of fundamental importance in this field is the development of external world modelling techniques, the fusion of data from different sensory modalities and the treatment of incomplete and/or uncertain information. These problems require the integration of 'high-level' capabilities (reasoning,

ARTIFICIAL VISION   Copyright © 1997 Academic Press Ltd
ISBN 0-12-444816-X   All rights of reproduction in any form reserved

planning, inferential activities) with 'low-level' capabilities (perception, motor control). To this aim, it appears promising to combine classic symbolic techniques of knowledge representation and reasoning with the computational techniques typical of connectionist models.

An approach based on connectionist architectures, like the one presented in this paper, represents an attempt to get over the well-known problems arising in order to describe simple objects as chairs and tables, by means of a first order logic. Moreover, the connectionist approach allows the integration of different knowledge representation levels via suitable mapping mechanisms, thus avoiding the need of an explicit control structure for the entire perceptual system, like a blackboard or other systems typical of artificial intelligence, which would imply a precise definition of the interaction mechanisms between levels. Another appealing feature of such hybrid models consists in the possibility of facing in a natural way high level problems, too hard to be solved within an entirely symbolic paradigm (e.g., uncertainty and incomplete information, prototypical representation of concepts, reasoning with analog models).

This paper deals with hybrid models of visual perception, intended to provide object recognition capabilities to an autonomous intelligent system acting in a unstructured environment. The conceptual description of scenes and the inferential activities at the symbolic level are grounded on the perceptual mechanisms by means of a connectionist mapping device. Visual perception is modelled as a process in which information and knowledge are represented and processed at different levels of abstraction, from the lowest one – directly related to features of perceived images – to the highest levels where the knowledge about the perceived objects is of a symbolic kind, through a 'conceptual level' where the geometric features of the scene are explicitly accounted for.

Therefore, it is a task of the higher level components to use the information acquired through the perceptual system, to create expectations or to form contexts in which the perceptual system's performance may be verified and, if necessary, modified by repeating the perceptual process once relevant parameters have been adequately changed (e.g. the placement of sensors, the detail level of 3D information extracted from sensory data, the characteristics of the resulting geometric representation, etc.). More in detail, we hypothesize three representation levels: the subsymbolic level, in which the information is strictly related to the sensory data; the linguistic level, in which information is expressed by a symbolic language; and an intermediate, 'prelinguistic' conceptual level. At this level the information is characterized in terms of a metric space defined by a certain number of 'cognitive' dimensions, independent from any specific language (Gärdenfors, 1992). The aim of this level is to generate the very internal representation of the agent's external environment and to provide a precise interpretation for the linguistic level.

The interpretation of linguistic level categories is implemented by a mapping between the conceptual and the linguistic levels in terms of a connectionist device. Neural networks allow to avoid an exhaustive description of conceptual categories at the symbolic level: in a certain sense, prototypes 'emerge' from the activity of an associative mechanism during a training phase based on examples. Moreover, the measure of similarity between a prototype and a given object is implicit in the behavior of the network and is determined during the learning phases.

A correlated cognitive aspect of hybrid architectures of this kind is the role of attentive processes in the link between the linguistic and the conceptual level. In fact, a finite agent with bounded resources cannot carry out a one shot, exhaustive, and uniform analysis of a perceived scene within reasonable time constraints. Furthermore, some aspects of a scene are more relevant then others, and it would be irrational to waste time and computational resources to detect true but useless details. These problems can be faced by hypothesizing a sequential attentive mechanism, that scans the internal representation of the scene.

The focus of attention is hypothesized to be driven on the basis of the knowledge, the hypotheses, the purposes and the expectations of the system, in order to detect the relevant aspects in the perceived scene. Hence, it is a task of the higher-level components to use the information acquired through the perceptual system to create expectations or to form contexts in which hypotheses may be verified and, if it is necessary, adjusted.

The proposed model must not be considered as a model of human vision: no hypotheses are made about this point, and the model may be referred to an autonomous abstract intelligent system, in which other components are devoted to the reasoning activities necessary for planning actions, controlling input sensors, coordinating motor activities, and so on.

The rest of this chapter is structured as follows: Section 9.2 outlines the design of a possible hybrid architecture for artificial vision, Sections 9.3 and 9.4 present the geometric primitives adopted and its application for solid modelling, Sections 9.5, 9.6 and 9.7 outline the computational steps from the image to the 3D reconstruction of the scene. Section 9.8 describes the conceptual level of representation, while Sections 9.9 and 9.10 respectively specify the linguistic level and its interpretation function. Section 9.11 particularizes in greater detail the focus of attention mechanism and Section 9.12 characterizes the links between the conceptual and the linguistic level in terms of time-delay attractor neural networks. Finally, Section 9.13 presents the main experimental results obtained with the implemented architecture.

9.2 THE DESIGN OF THE ARCHITECTURE

The cognitive assumptions introduced in the previous section are the guidelines for the design and implementation of an architecture for artificial vision (Chella

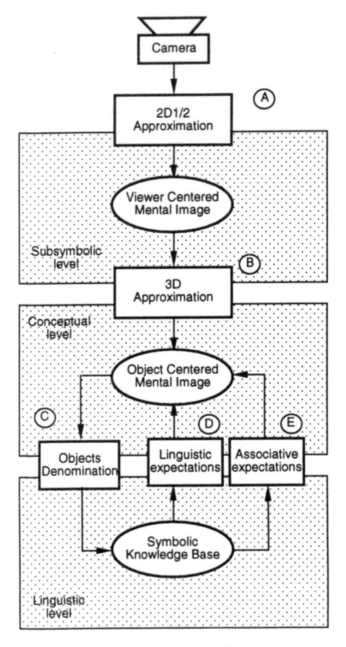

Figure 9.1 The proposed architecture. The three levels of representation are evidenced.
Block A receives the input from a camera and gives as an output the $2\frac{1}{2}D$ depth images.
The maps are sent to block B, which builds a scene description in terms of a combination
of 3D geometric primitives. Block C implements the associative mapping between the
conceptual level and the symbolic one. Block D implements the linguistic modality of the
focus-of-attention mechanism, while block E implements the associative modality of
the focus of attention.

et al., 1994). Figure 9.1 shows the overall architecture in which the previously described three levels of representations are evidentiated.

Block A is the kernel of the subsymbolic level: it receives one or more input pictorial digitized images acquired by a camera and, by means of shape from shading algorithms, gives as an output the $2\frac{1}{2}$D depth images. The $2\frac{1}{2}$D depth images are input to block B, which builds a scene description in terms of a combination of 3D geometric primitives. Block C implements the associative mapping between the conceptual level and the symbolic level; the aims of this block is to recognize the objects and situations. The input to block C is a vector configuration in the conceptual space, its output is sent to the linguistic level to produce a symbolic description of the scene.

The symbolic knowledge base represents the kernel of the linguistic level. The aim of this block is twofold: it describes in a high-level language the perceived scene by interpreting the input coming from block C, and generates, by means of its inference capabilities, the linguistic 'expectations' that drive the focus-of-attention mechanism.

Three focus-of-attention modalities are hypothesized as the basis of the proposed architecture: a reactive modality, in which the attention is driven only by the characteristics of scene, a linguistic modality in which the attention is driven by simple inferences at the linguistic level and an associative modality in which the attention is driven by free associations among concepts.

Block D is the block responsible for the linguistic modality of the focus of attention mechanism. It receives as input the instances of concepts from the knowledge base and it suitably drives the focus of attention, in order to seek for the corresponding objects and situations in the acquired scene. Block E is responsible for the associative modality of the focus of attention. Its operation is similar to block D, but it drives the focus of attention looking for the objects in the scene which can be freely associated to the input instances.

The reactive modality of the focus of attention is implemented as an internal mechanism of block D: when the block does not receive any expectations as input, it generates some generic expectations in order to start the operations of the system.

9.3 THE GEOMETRIC PRIMITIVES

According to the classic approach by Marr and Nishihara (1978), the first step towards the recognition of objects present in a scene requires to derive a 3D model description from a scene image, through the identification, by information extracted from the image, of the model's coordinate system, of the relative spatial arrangement and sizes of major parts of the model, and of shape components (the *primitives* representing the low-level 'parts' of the description). This organization guarantees that the description can capture the geometry of a shape to an arbitrary level of detail, and that none of the higher-level processes using the representation for recognition has to deal with the internal details of more than one 3D model at a time, even if the complete description of a shape involves many 3D models. Obviously, representations of this kind are strongly influenced by the choice of primitive elements and of composition operators. The original model of Marr and Nishihara is based on hierarchies of *generalized cones*, so its scope is limited to shapes for which the determination of natural axes is based on a shape's elongation. Therefore, such a representation is capable only of a highly abstracted description of most of natural forms.

Other types of engineering representations based on descriptions of the object's surface shape have been presented in the areas of computer graphics, CAD/CAM, and so on. *CSG* (Constructive Solid Geometry) *representations*, for example, describe complicated solids as various 'additions' and 'subtractions' of simpler solids by means of modified versions of Boolean operators – union, difference and intersection – e.g., regularized operators (Requicha, 1980). *Boundary* and *sweeping* representations also allow for unambiguous representations of solids, but are characterized by a limited scope, in that mathematical and computational issues become considerably complex when the shapes to be represented lack the 'regularity' that characterizes man-made objects. Therefore, the latter schemes seem to be inadequate in order to more simply and concisely represent natural and biological forms (people, mountains, clouds, etc.), as required of a modelling scheme to be used in a general-purpose vision system.

In (Pentland, 1986) it is proposed an effective representation system for the description of both natural and man-made forms, where the representation primitives, i.e., the building blocks of complicated objects, are modelled by using superquadrics. Superquadrics are mathematical shapes based on the parametric form of quadric surfaces, in which each trigonometric function is raised to one of two real exponents (Barr, 1981). Thus a superquadric is completely defined by: its center, orientation, 'sides' (length, width and breadth) and the above-mentioned real exponents, ε_1 and ε_2, called *form parameters* since a change in these exponents affects the superquadric's surface shape. For example, superellipsoids in canonical positions are described by the following equations ($C_\eta = \cos \eta$, $S_\omega = \sin \omega$, and so on):

– *Position* and *normal* vectors of the surface, respectively:

$$x(\eta, \omega) = \begin{bmatrix} a_1 C_\eta^{\varepsilon_1} C_\omega^{\varepsilon_2} \\ a_2 C_\eta^{\varepsilon_1} S_\omega^{\varepsilon_2} \\ a_3 S_\eta^{\varepsilon_1} \end{bmatrix} \tag{9.1}$$

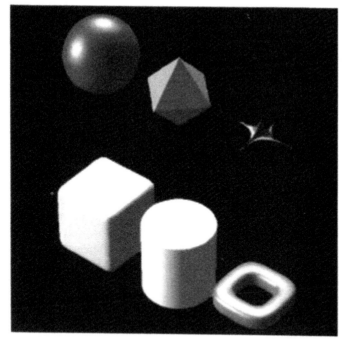

Figure 9.2 Examples of superquadrics. Values of shape factors near to 0 give squared shapes, while increasing the shape factors rounded and pinched shapes are generated.

$$n(\eta, \omega) = \begin{bmatrix} \dfrac{1}{a_1} C_\eta^{2-\varepsilon_1} C_\omega^{2-\varepsilon_2} \\[2mm] \dfrac{1}{a_2} C_\eta^{2-\varepsilon_1} S_\omega^{2-\varepsilon_2} \\[2mm] \dfrac{1}{a_3} S_\eta^{2-\varepsilon_1} \end{bmatrix} \qquad (9.2)$$

- *Inside–outside* function:

$$F(x, y, z) = \left(\left(\frac{x}{a_1}\right)^{2/\varepsilon_2} + \left(\frac{y}{a_2}\right)^{2/\varepsilon_2} \right)^{\varepsilon_2/\varepsilon_1} + \left(\frac{z}{a_3}\right)^{2/\varepsilon_1} \qquad (9.3)$$

where $-\pi/2 \leqslant \eta \leqslant \pi/2$, $-\pi \leqslant \omega < \pi$ and a_1, a_2, a_3 are respectively the length, width and breadth (the 'sides') of the superquadric.

The surfaces so described are parameterized in latitude η and longitude ω, and their shape is controlled by the parameters ε_1 and ε_2, in latitude and longitude respectively. Modifying shape factors causes a local surface deformation that can result in squared as well as pinched or rounded shapes (Figure 9.2). In this way, the simple modification of a few parameters allows for a number of different shapes. Global deformations can also be defined, to take into account stretched, twisted and bent shapes (Ardizzone *et al.*, 1989; Solina and Bajcsy,

1990). Thus, Boolean combinations and deformations of superquadrics are able to represent with surprising realism scenes and objects of the real world. On the basis of arguments of this kind, superquadrics have received in the last years considerable attention from computer vision scientists.

9.4   THE SOLID MODELLING SYSTEM

The solid modelling scheme falls in the area of CSG modelling systems: a complex 3D shape can be described by a tree, in which the leaves represent primitive shapes (superquadrics) and the internal nodes represent either the shape resulting from the application to an input shape of *unary* operators (deformations, spatial rigid movements, scaling operations) or the shape resulting from the application of *binary* (Boolean difference) or *n-ary* (Boolean union and intersection) operators. In the former case the node has only one offspring, in the latter one, two or more offspring. Complex operations however can always be reduced to this conceptually simple scheme, by introducing, if necessary, intermediate shapes and operations, thus following a step-by-step approach to the modelling of complex solids.

In addition to the regularized Boolean operators '·' (intersection), '+' (union) and '−' (difference) (see below), the modelling algebra comprises the unary deformation operators ST (stretching), BN (bending) and TW (twisting). Geometric movements like translation and rotation, along with scaling operations, can be taken into account by the unary *position-scale* operators PS, which simply applies a rototranslation and scaling matrix to the input shape. Table 9.1 summarizes the operators present in our modelling system.

Boolean operations

Following CSG schemes, complex objects can be obtained as recursive, Boolean combinations of representation primitives. Objects and Boolean operators are assumed to be *regularized*, that is objects cannot have degenerate or overlapping boundaries, and the results of operations produce other regularized objects (Tilove, 1980).

Table 9.1

·	Boolean intersection of input shapes
+	Boolean union of input shapes
−	Boolean difference of input shapes
ST	Stretching of input shape
BN	Bending of input shape
TW	Twisting of input shape
PS	Positioning and scaling of input shape

Given two objects **A** and **B**, the Boolean operations of *intersection* $(\mathbf{A} \cdot \mathbf{B})$, *union* $(\mathbf{A} + \mathbf{B})$ and *difference* $(\mathbf{A} - \mathbf{B})$ can be considered in terms of parts of **A**s and **B**s boundaries (Putnam and Subrahmanyam, 1986). In particular, the boundary portions of **A** and **B** can be classified as belonging to the following sets:

– AinB	parts of the boundary of **A** that lie inside **B**
– BinA	parts of the boundary of **B** that lie inside **A**
– AoutB	parts of the boundary of **A** that lie outside **B**
– BoutA	parts of the boundary of **B** that lie outside **A**
– ABshared	parts of the boundaries of **A** and **B** that are coincident and bound the same region
– ABanti_shared	parts of the boundaries of **A** and **B** that are coincident and bound disjoint regions.

The boundary of the shape resulting from one of Boolean operations applied to **A** and **B** is given by:

$$\mathbf{A} + \mathbf{B} = \text{ABshared} + \text{AoutB} + \text{BoutA}$$

$$\mathbf{A} \cdot \mathbf{B} = \text{ABshared} + \text{AinB} + \text{BinA} \qquad (9.4)$$

$$\mathbf{A} - \mathbf{B} = \text{ABanti\_shared} + \text{BinA} + \text{AoutB}.$$

Therefore, implementing Boolean operations requires the partitioning of **A**s and **B**s boundaries and subsequently the classification of the obtained boundary portions. To this aim, it is necessary, first of all, to evaluate the boundaries starting from the CSG representation of solids, that are in general non-primitive solids. Moreover, the result of the Boolean operation is of the same form as the operands, i.e., a boundary representation. Efficient and reliable algorithms of conversion between CSG and boundary representation, and vice versa, are used to maintain the consistency of the modelling scheme.

As far as the classification of boundary parts is concerned to meet the superquadric equations, an algorithm based on the inside-outside function can be used to determine if an arbitrary point falls at the inside or at the outside of the solid, or on the boundary. For example, the unit sphere has the inside-outside function:

$$f(x, y, z) = x^2 + y^2 + z^2$$

and therefore we have:

If $f(x_0, y_0, z_0) = 1$, (x_0, y_0, z_0) is on the surface.
If $f(x_0, y_0, z_0) > 1$, (x_0, y_0, z_0) lies outside the sphere.
If $f(x_0, y_0, z_0) < 1$, (x_0, y_0, z_0) lies inside the sphere.

The bounding surface is *not* included in the solid description because of regularization. Once calculated AinB, etc. the object resulting from the Boolean combination of primitives can be immediately calculated by eqn 9.4.

The Boolean difference operator is always applied to pairs of input shapes, conceptually defining a single object as output shape; on the contrary, the Boolean union and intersection can have a number of component shapes, conceptually defining either a single object or an entire scene as output.

Deformations

Superquadrics may be considered the basic prototypes of our modelling system; in addition to Boolean combinations, global deformations can also be taken into account in order to enlarge the scope of the representation scheme. Thus stretching, bending and twisting of primitives can be used to form new, more complex prototypes. Every operation of this type has only one input and one output shape.

More complex deformations, i.e., combined deformations, must be reduced to the latter simple scheme.

Some basic questions are:

a) the definition of deformation operations, in terms of *parameters* characterizing them and *operators* implementing them;
b) the order of subsequent deformations, if more than one is to be applied.

The latter point is related to the more general problem of describing complex objects, and will not be further discussed in this paper.

As far as the former point is concerned, if deformed primitives have to be used for shape recovery from sensory data, they must retain an analytic and parametric formulation, and must be invertible. Keeping this in mind, given a superquadric in canonical position, global deformations of bending, twisting and stretching, as formally shown in Ardizzone and Ardizzone (1995) and Ardizzone *et al.* (1995), can be defined in the following way.

Bending
a) Only one of the superquadric principal axes, called deformation axis, is deformed from a straight line to a curve, and the corresponding side retains its length.
b) Cross sections perpendicular to the deformation axis are assumed to be invariant in form and size during the deformation, and perpendicular to the deformed axis also after the deformation.

Twisting
a) Only one of the superquadric principal axes is twisted, and the corresponding side retains its length.
b) Cross sections perpendicular to the deformation axis are assumed to be invariant in form and size during the deformation, and perpendicular to the deformed axis also after the deformation.

Stretching

a) Only one of the superquadric principal axes is stretched, and the corresponding side is expanded or contracted, depending on the (positive or negative) stretching coefficient.

b) Cross sections perpendicular to the deformation axis are assumed to be invariant in form, but their size is scaled during the deformation, to fill the volume conservation constraint.

9.5 FROM 2D DATA TO THE 3D GEOMETRIC REPRESENTATION OF OBJECTS

The fitting between superquadrics and single parts of 3D information (range data or 3D points extracted from 2D image data) has been addressed by many authors. A review of reconstruction methods using either superquadrics or other forms of 3D surface representations may be found in Bolle and Vemuri (1991). Several approaches are taken into consideration in the following.

Starting with the observation that image data is mostly a function of surface normals, and from the duality relation between surface normal and surface shape of superquadrics, in Pentland (1987) it is shown that overconstrained estimates of a superquadric's parameters can be obtained from image data. Closed form solutions, in particular, can be obtained for the 11 parameters characterizing a superquadric in general position, having information from 2D contours and shading as input and using linear regression to compute parameter values providing the best fit. Pentland yet approaches the recovery of superquadrics from sparse range data, by making use of heuristics to compute a 'goodness-of-fit' functional in a search over the entire parameter space.

Solina and Bajcsy (1990) address the problem of superquadric recovery from 3D points available from a range imager, as a least-squares minimization problem. Their iterative fitting procedure is based on the definition of a cost function whose value depends on the distance of points from the superquadric's surface and on the overall size of the superquadric. The inside-outside function rewritten for a superquadric in general position and orientation in the world coordinate system is used. In particular, given N 3D points $(x_{w_i}, y_{w_i}, z_{w_i})$, $i = 1, \ldots, N$, they minimize the expression:

$$\sum_{i=1}^{N} [R(x_{w_i}, y_{w_i}, z_{w_i}; a_1, \ldots, a_{11})]^2 \tag{9.5}$$

instead of

$$\sum_{i=1}^{N} [1 - F(x_{w_i}, y_{w_i}, z_{w_i}; a_1, \ldots, a_{11})]^2 \tag{9.6}$$

and *R* is the functional:

$$R = \sqrt{a_1 a_2 a_3}(F - 1) \qquad (9.7)$$

where $F = F(x_w, y_w, z_w; a_1, \ldots, a_{11})$ is the inside-outside function, and $a_1, \ldots, a_{11}$ are the 11 independent parameters defining the superquadric – center position (3 parameters), object orientation (3 parameters), size (3 parameters) and form parameters (2 parameters). The adoption of eqn (9.7) allows the use of a function with a minimum corresponding to the *smallest* superquadric, fitting the given set of 3D points and such that the function value for surface points is known *before* minimization. Superquadric sides and form parameters are constrained: $a_1, a_2, a_3 > 0$ and $0.1 < (\varepsilon_1, \varepsilon_2) < 2$.

Given initial trial values for $a_1, \ldots, a_{11}$, the procedure is repeated until the sum of least squares stops decreasing, or the changes are statistically meaningless. The choice of the set of initial values is based on a rough computation of position, orientation and size of the superquadric. The recovery of deformed superquadrics is also possible, by taking into account in the previous expression of the inside-outside function also the parameters defining the deformation.

A completely different approach has been followed by Dickinson *et al.* (1992). A limited set of pure and deformed superquadrics is used to generate a hierarchical aspect representation, based on the projected surfaces of primitives; conditional probabilities capture the ambiguity of mappings between hierarchy levels. The input image is segmented, and regions are grouped into aspects. No domain-independent heuristics are used; only the probabilities inherent in the aspect hierarchy are used. Once the aspects are recovered, the aspect hierarchy is employed to infer a set of volumetric primitives and their connectivity.

Our approach to the primitive recovery from image data directly follows the common notion of 'parts' and then takes a straightforward advantage of the CSG nature of the proposed modelling system. Since we are concerned with gray-scale images, early vision methods, like shape-from-x algorithms, can be used to reconstruct the depth information, using one or more 2D images of the observed scene, in order to build a 3D shape representation. Shape parts are subsequently approximated by superquadrics. No 3D shape segmentation is currently attempted: when necessary, 2D images are directly segmented in order to obtain scene portions to be approximated by single superquadrics.

This heavy limitation, that will be removed when effective 3D segmentation algorithms will be available, may be partially justified by the simplicity of scenes currently considered. In each case, connectivity and other relations among scene parts are taken into account at the conceptual level of the vision system.

In the rest of the paper, experimental results will be shown with reference to the simple real scene, made up by a hammer, a tennis ball and a computer mouse put on a table, shown in Figure 9.3. The scene has been acquired as a 256×256, 8 bits per pixel image.

9.6 FROM THE IMAGES TO THE SHAPE

We need the extraction from 2D images of 3D information adequate to the subsequent fitting with superquadrics. We are not really interested, in this phase of the recovery procedure, to the direct determination of closed form solutions for the surface. A good, but not necessarily exact estimate of the objects' shape may be sufficient, since the recognition and classification problems are managed at the linguistic level. This may be achieved if depth values of the entire image, i.e. a sort of *dense* depth map, are available.

Robust shape-from-X algorithms are available for this task. Shape-from-shading and shape-from-stereo are among the most popular approaches (Horn, 1986), even if their performance is normally acceptable only in the presence of limitative hypotheses. The most important of these limitations are on the surface form and characteristics, and on the environmental control of illumination sources. Some of these limitations can be at present accepted in our application. Shapes may be supposed convex, as required by the fitting procedure and by the used geometric primitives, and the surface must be sufficiently smooth and homogenous; moreover the environment in which the images are acquired must be sufficiently controlled.

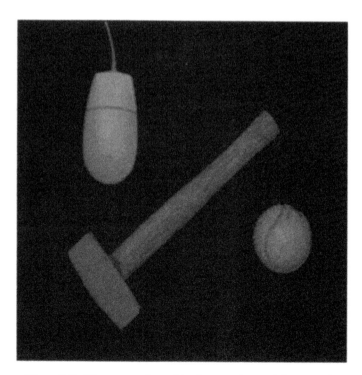

Figure 9.3 The scene adopted in the experimental framework.

Shape-from-shading algorithms are normally subdivided into two general classes. In *global* methods, the shape is recovered by minimizing some cost function involving constraints such as smoothness. The shape is iteratively computed, e.g., using variational techniques (Horn and Brooks, 1989), and it maintains itself globally consistent. In *local* methods, the estimate of the shape is attempted from local variations in image intensity, by using local constraints about the surface or about the reflectance map. Local methods are simpler and less accurate, while global methods are more complex and more precise.

On the basis of the above considerations, we use two methods originally proposed by Pentland, respectively known as *local* shape-from-shading (Pentland, 1989) and *linear* shape-from-shading (Pentland, 1990). In both cases, the surface is assumed to be locally spherical, and the Lambertian reflectance model is adopted, with constant albedo. In the second method, a linear approximation to the true reflectance function is used. The illuminant direction is computed along with the shape information by the local algorithm, while the linear algorithm requires a separate computation.

Another method, similar to the Pentland's linear algorithm, has been also taken into consideration, i.e. the algorithm proposed by Tsai and Shah (1992). The major difference lays in the fact that Pentland uses a reflectance map linear in the surface gradient (p, q), while Tsai and Shah linearize the reflectance in the depth $Z(x, y)$.

These three approaches have been implemented and used on the image shown in Figure 9.3. The results are shown in Figure 9.4, for Tsai and Shah's method. To give the users an adequate impression of the estimated shapes, a spline interpolation, available in our graphic system, has been adopted for the visualization.

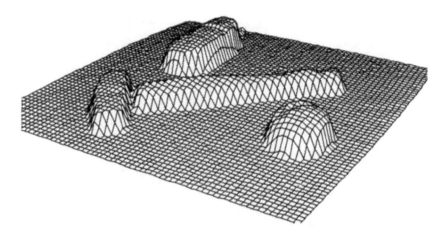

Figure 9.4 The shape reconstructed by using the Tsai and Shah's method.

9.7   FROM THE SHAPE TO THE SUPERQUADRICS

As the dense depth map of the scene under consideration has been computed, the next step in the visual process is the reconstruction of the geometric model, i.e., the recovery of the superquadrics that best fit the various scene parts.

The implemented fitting procedure uses the same functional (eqn 9.5) adopted by Solina and Bajcsy (1990). Major differences are:

i)   We address the minimization problem by applying a conjugate gradient method instead of a least-squares fitting, in order to provide a sufficiently fast goodness-of-fit and a robust behavior with respect to local minima.

ii) Our initial data is the $2\frac{1}{2}$ D sketch, i.e. a dense depth map of the scene under consideration, instead of laser range-finder data.

As far as the point (ii) is concerned, the use of true 3D points allows for a choice of initial parameter values of the following kind. Initial values for shape parameters ε_1 and ε_2 are always 1, meaning that the initial model is always an ellipsoid. The superquadric's center is assumed to be initially coinciding with the center of gravity of all N range points, i.e. the center's coordinates are determined as mean values (in x, y and z) of the available points' coordinates. The directions of the three axes of the object centred reference system are calculated from the eigenvectors of the matrix of central moments. Axes are labelled so that the axis z lies along the longest side for elongated objects or perpendicular to flat or rotationally symmetric objects. The directions are expressed in terms of Euler angles. Finally, sides of the initial ellipsoid are simply computed by the distances between the outermost range points along each found direction.

Results of the minimization procedure by using the so found initial values are described in the subsequent Section 9.13.

9.8   THE CONCEPTUAL LEVEL OF REPRESENTATION

In Marr's theory a superior symbolic level is limited to a hierarchically organized catalogue of 3D prototypes. The mental imagery literature (Glasgow, 1993) hypothesizes a distinction between mental pictures and propositional mental representations; for example, Kosslyn (1980) distinguishes between a short-term memory based on mental images, and a propositional long-term memory. Cognitive evidence exists, according to which both these kinds of representation coexist and are integrated in the human memory (Farah *et al.*, 1988). In the theory of Johnson-Laird (1983), three levels of representation are hypothesized, that, in a certain sense, summarize the various points of view sketched above.

From a slightly different point of view, Gärdenfors (1993) proposes three levels of information representation: linguistic, conceptual and subsymbolic. At the linguistic level the information is described in terms of a symbolic language, e.g., first order logic; at the subsymbolic level, information is characterized directly in terms of the perceptual inputs of the system. Between these two levels, a third level is hypothesized: the conceptual one, in which information is described by means of a conceptual space.

Our model is inspired to the Gärdenfors' proposal. Many analogies can be found in the Marr model and in many positions emerged from the mental imagery debate. In Figure 9.1, the three gray blocks correspond to the Gärdenfors' levels of representation. The first level can be also regarded as a visual, viewer-centered mental image (or, in the Marr's terminology, $2\frac{1}{2}$D sketch). The central level embeds an object-centered mental image (in the Marr's terminology, a 3D model representation). The upper level consists of a propositional, linguistic knowledge representation. Such a level can be assimilated to the Kosslyn's long-term memory and to Marr's hierarchical catalogue of models.

A conceptual space is a metric space consisting of a number of quality dimensions. From a formal point of view, a conceptual space is an n-dimensional space CS where X_i is the set of values of the ith quality dimension ($1 \leqslant i \leqslant n$). Examples of such dimensions can be color, pitch, mass, spatial coordinates, and so on. The dimensions should be considered 'cognitive' in that they correspond to qualities of the represented environment, without references to any linguistic descriptions. In the case of visual perception, the dimensions of the conceptual space correspond to the parameters of a suitable system of geometric primitives, e.g. simple blocks (as Marr's generalized cylinders), geons (Biederman, 1985) and superquadrics.

We define the *knoxel* as a generic point in a conceptual space (the term 'knoxel' is suggested by the analogy with 'pixel'); knoxels therefore represent the epistemological primitive elements at the considered level of analysis. Formally, a knoxel is a vector $k = \langle x_1, x_2, \ldots, x_n \rangle$ where $x_i \in X_i$ ($1 \leqslant i \leqslant n$); each component corresponds to a parameter describing a quality dimension of the domain of interest. In the case of visual perception, these parameters characterize the geometric primitives of the kind quoted above. In this perspective, the knoxels correspond to simple geometric building blocks, while complex objects or situations are represented as suitable sets of knoxels. Accordingly, each knoxel is related to measurements, obtained via suitable sensors, of the geometric parameters of 'simple basic' objects in the external environment.

A metric function d is defined in CS, which may be considered as a measure of similarity among knoxels in the conceptual space.

For the sake of simplicity we deal with a countable set of knoxels by discretizing the axes of the conceptual space CS. As a result the set of all perceivable knoxels k_i is the countable set $KS = \{k_1, k_2, \ldots\}$.

As previously explained, perceived objects and situations correspond to suitable sets of knoxels; we define a perception cluster $pc = \{k_1, k_2, \ldots, k_l\}$ on a finite set of knoxels corresponding to an object or a situation in KS. The set PC of all the perception clusters in KS is defined as:

$$PC = \{\{k_1, k_2, \ldots, k_l\} \mid l \in N, k_i \in KS \text{ for } 1 \leqslant i \leqslant l\}. \qquad (9.8)$$

The conceptual level is independent from any linguistic characterization. On the contrary, the semantics of the symbols at the linguistic level is based on the conceptual level, since symbols are interpreted on configurations at the conceptual level, as described in Section 9.10.

9.9 THE LINGUISTIC LEVEL

The role of the linguistic level in the proposed architecture is to provide a rough and concise description of the perceived scene in terms of a high-level logical language, suitable for symbolic knowledge-based reasoning.

In order to describe the symbolic knowledge base we adopt a hybrid representation formalism, in the sense of Nebel (1990). According to this point of view, a hybrid formalism is constituted by two different modules: a terminological component and an assertional component. In our model, the terminological component contains the descriptions of the concepts relevant for the represented domain (e.g., types of objects and of situations to be perceived). The assertional component stores the assertions describing the specific perceived scenes.

The choice of a hybrid formalism is not constraining. The distinction between terminological and assertional components, however, can be useful to keep distinct the conceptual knowledge, which is largely independent from the specific perceived scene, and the assertions concerning the scene itself. In this sense, the terminological component can be seen as analogous to a long-term memory, while the assertional component to a short-term memory. Moreover, terminological formalisms are well suited for our purposes, in that they are centered on conceptual descriptions. This allows a compact description of concepts, whose instances are to be recognized in the perceived scene.

As an example, consider in Figure 9.5 a simple fragment of the terminological knowledge base, concerning the description of objects. In the figure, the network notation developed for the KL-ONE system (Brachman and Schmolze, 1985) has been adopted. A brief description of the KL-ONE formalism may be found in Appendix 9.A.

The assertional component is based on a first order predicate language, in which the concepts of the terminological component correspond to one argument predicates, and the roles (e.g., *hammer head* or *hammer handle*) correspond to two argument relations.

Concerning situations, they are also represented as concepts in the terminological formalism. In other words, we assume that situations are reified, i.e., that

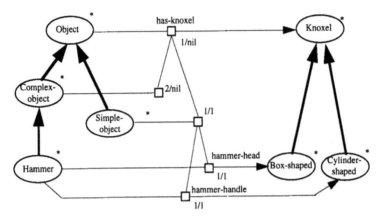

Figure 9.5 A fragment of the terminological knowledge base. A generic *object* is described as composed by at least one knoxel. A *simple object* is described as an object composed by exactly one knoxel; a *complex object* is an object approximable by at least two knoxels. *Hammer* is an example of complex object. The role *has-knoxel* has been differentiated in more distinct roles. The concept *hammer* has two roles: a role *hammer handle* and a role *hammer head*.

to every specific situation corresponds an individual in the domain. This gives a great flexibility and expressive power. Figure 9.6 shows the network description of the *situation* concept, and of two particular types of situation, *on* and *sided*.

9.10 THE MAPPING BETWEEN THE CONCEPTUAL AND LINGUISTIC LEVELS

The mapping between the conceptual and the linguistic levels defines an internal, cognitively-oriented, semantic interpretation for the symbols at the

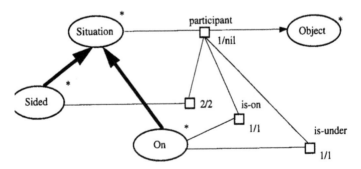

Figure 9.6 Network description of the *situation* concept. Every situation has at least one object as participant. *Sided* is described as a particular type of situation, with exactly two participants.

linguistic level. The interpretation function associates a perception cluster to any individual constant representing an object or a situation at the linguistic level a perception cluster. Therefore, if C is the set of assertional constants representing objects and situations, the interpretation function φ has the following type:

$$\varphi: C \to PC \tag{9.9}$$

where PC represents the set of all perception clusters as defined in eqn 9.8.

The assertional language can also treat assertional constants, representing more specific features of objects and situations, like, for instance, the main axis of an object or its shape. The set C' of such assertional constants is mapped by the interpretation function to the specific structures of the conceptual space:

$$\varphi: C' \to X \tag{9.10}$$

where $X = \bigcup_{i=1}^{n} X_i$ is the set of all the values of the quality dimensions of the conceptual space.

The compositional aspects of the interpretation of symbolic structures at the linguistic level can be defined according to the usual model-theoretic semantics of terminological languages.

The main difference between the proposed internal semantics and the usual model-theoretic semantics is that the extensions of predicates are uniquely defined at the conceptual level and, therefore, the truth of atomic assertions in which primitive concepts are involved can be easily determined by specific relations among the conceptual entities.

Consider, for example, a *participant* role. Given the assertion *participant* (i, j), in a purely extensional model-theoretic semantics its truth is justified solely by the fact that the pair of the extensions of i and j belongs to the extension of *participant*:

$$\langle \varphi[i], \varphi[j] \rangle \in \varphi[\,participant\,] \tag{9.11}$$

In the internal semantics, the truth of *participant*(i, j) can be determined by examining the entities on which i and j are interpreted in the conceptual space: *participant*(i, j) is true if the set of knoxels on which j is interpreted is a subset of the set of knoxels on which i is interpreted:

$$\varphi[j] \subset \varphi[i] \tag{9.12}$$

9.11 THE FOCUS-OF-ATTENTION PROCESS

As sketched in the introduction, a finite agent with bounded resources cannot carry out a one-shot, exhaustive, and uniform analysis of a perceived scene within reasonable time constraints.

In modelling perception, these problems can be faced taking into account the fundamental role of attentive phenomena in vision (Yarbus, 1967). In the psychological literature, the focus of attention has been sometimes described as a light spot which scans the visual field, individuating relevant aspects (Posner, 1980). This mechanism is analogous to the scanning of a mental image, as described by Kosslyn (1980). Several models of focus-of-attention mechanisms have been proposed in the active vision literature. The interest in this field has been summarized by Bajcsy and Campos (1992) who propose the 'active and exploratory' framework for perception. A largely used strategy to model the focus of attention, followed for example by Burt (1988), is based on the pyramidal approach. Rimey and Brown (1994) present TEA-1, a task-oriented system performing the minimum effort necessary to solve a specific task. Birnbaum et al. (1993) propose the BUSTER system, aimed at developing a causal explanation of the scene.

In our architecture, the conceptual level described in the previous section acts as a 'buffer interface' between subsymbolic and linguistic processing. The information coming up from the subsymbolic level has the effect of contemporary activating an (eventually very large) set of knoxels in the conceptual space. It is the focus of attention mechanism that imposes a sequential order in the conceptual space according to which the linguistic expressions can be given their interpretation.

In order to model the focus of attention mechanism, we define a *perception act* p as a generic sequence of knoxels in the conceptual space:

$$p \in KS^* \tag{9.13}$$

where KS^* is the set of all sequences of knoxels in KS.

With reference to a perception cluster pc, we say that a perception act p is associated to the perception cluster pc if $p \in pc^*$, where pc^* is the set of all sequences of knoxels belonging to pc. The perception act associated to a perception cluster therefore corresponds to a specific way of perceiving an object or situation described by the perception cluster. It should be noted that the sequence of knoxels making a perception act may not include all the knoxels of the corresponding perception cluster and/or it may include several times the same knoxels.

In this context we introduce the denotation function ϑ mapping a perception act to the corresponding assertional constant:

$$\vartheta: KS^* \to C \tag{9.14}$$

$\vartheta(p)$ is therefore the assertional constant perceived by the perception act p. Block C in Figure 9.1 implements the denotation function by means of a suitable neural network, as described in the next section.

The description of complex elements in terms of sequences of knoxels seems to be a natural extension of the Gärdenfors' notion of conceptual space. It should be noted that the perception act assumption avoids the necessity to augment the dimensions of the space in order to describe complex objects or situations made up by several blocks: they are described by perception acts of several length.

In order to individuate the grouping paths among knoxels and generate the most meaningful perception acts, it is necessary to suitably orient the focus of attention. In the proposed architecture the focus of attention is determined by three concurrent modalities: the reactive, the associative and the linguistic modality. The 'reactive' modality is the simpler one: the grouping paths among knoxels are determined only by the characteristics of the visual stimulus, e.g. the volumetric extension of the forms, or the aggregation density of the perceived objects.

In the associative modality, the grouping paths are determined by an associative, purely Hebbian mechanism determining the attention on the basis of free associations between concepts. Whenever two objects in the same scene are perceived, the weight of the associative connection between the corresponding concepts is increased.

In the linguistic modality, the focus of attention is driven by the symbolic information explicitly represented at the linguistic level. Consider the hammer example: at the linguistic level (see Figure 9.5) a hammer is described as composed by a handle and a head. If an object similar to a hammer handle has been recognized, the linguistic level hypothesize that a hammer may be present in the scene. The focus of attention is directed to try to identify its parts, in particular its handle and its head, in order to confirm the presence of such an hammer in the scene. This corresponds to find the suitable fillers for the role parts of the object, i.e., a filler for the *hammer-head* role and a filler for the *hammer-handle* role. Therefore, whenever a hammer handle is recognized, the focus of attention tries to identify a hammer by identifying suitable fillers for its head and its handle.

The focus of attention mechanism may be modelled as an expectation function ψ linking the linguistic to the conceptual level; the function has its domain in the set C of assertional constants representing the expected objects or situation and its range in the set of perception acts belonging to the corresponding perception clusters. In other worlds, the focus of attention looks for specific perception acts belonging to the perception clusters corresponding to the 'expected' assertional constant in the perceived scene. The function ψ has the following type:

$$\psi^i : C \to KS^* \qquad i = 1, 2 \tag{9.15}$$

where KS^* is the set of all perception acts; i indicates the attentive modality: 1 stands for the linguistic modality, and 2 for the associative one.

In the linguistic modality the assertional constants grow up from inferences in the linguistic formalism; the focus of attention generates perception acts made up by knoxel samples for these assertional constants.

According to the hammer example, when a hammer is found, the functions:

$$\psi^1[Hammer - head - filler\#1] \text{ and } \psi^1[Hammer - handle - filler\#1]$$

generate perception acts made up by sample fillers of the role *hammer head* and *hammer handle*.

The associative modality is similar to the linguistic modality, except that the focus of attention searches for objects freely associated to the constants growing up from the linguistic formalism. According to the hammer example, when a hammer is found, the function $\psi^2[Hammer\#1]$ generate perception acts made up by samples of balls.

In the 'reactive' modality the focus of attention searches for generic objects in the scene; for uniformity, this modality may be considered as a special case of the linguistic modality where the expected assertional constant is an istantiation of the most generic class, e.g. Object#1.

9.12   THE NEURAL NETWORK IMPLEMENTATION OF THE MAPPING BETWEEN THE CONCEPTUAL AND THE LINGUISTIC LEVELS

This section outlines the neural network implementation of the mapping between conceptual and linguistic level. A more detailed presentation may be found in Chella *et al.* (1994).

A perception cluster, as previously described, is a set of knoxels associated to a perceived object or situation: $pc = \{k_1, k_2, \ldots, k_n\}$. Each knoxel k_i may be viewed as a point attractor of a suitable energy function associated to the perception cluster.

A set of fixed point attractors is a good candidate as the model for a perception cluster: starting from an initial state representing a knoxel imposed, for instance, from the external input, the system state trajectory is attracted to the nearest stored knoxel of the perception cluster. Therefore, the implementation of perception clusters by means of an attractor neural network (Hopfield, 1982) appears natural. Following this approach the implementation of the perception acts associated to a perception cluster is built by introducing time delayed connections storing the corresponding temporal sequences of knoxels. It should be pointed out that the choice of the time-delayed attractor neural networks is not constraining but offers several advantages. It is based on the well-studied energetic approach; the learning phase is fast, since it is performed at 'one shot'. Furthermore, as it allows for a uniform treatment of recognition and generation of perception acts, the denotation functions and the expectation functions introduced in the previous section may be implemented by a uniform neural network architecture design.

In order to describe the dynamics in the conceptual space an adiabatically varying energy landscape E is defined. The energy E is the superimposition of three energies (eqn 9.16): E_1 represents the fast dynamics for period of duration t and it models the point attractors for the single knoxels belonging to the perception clusters; E_2 represents the slow dynamics for period of duration $t \gg t_d$ due to time-delayed connections and it models the perceptions acts; E_3 model the global external input to the network.

The global energy function of the time-delayed synapses attractor neural network is (Kleinfeld, 1986):

$$E(t) = E_1(t) + \lambda E_2(t) + \varepsilon E_3(t) \qquad (9.16)$$

where E_1, E_2, E_3, are the previously described energy terms; λ and ε are the weighting parameters respectively of the time delayed synapses and the external input synapses.

The expectation functions ψ^i describing blocks D and E of our architecture are implemented by setting of parameters of the energy function E to $\lambda > 1$ and $\varepsilon = 0$. In fact, the task of these blocks is the generation of suitable knoxel sequences representing the expected perception acts.

This choice of parameters allows the transitions to occur 'spontaneously' with no external input. Referring to eqn (9.16), an attractor is stable for a time period significantly long due to the E_1 term. As $\lambda > 1$, the term λE_2 is able to destabilize the attractor and to carry the state of the network toward the successive attractor of the sequence representing the successive knoxel of the stored perception act. The neural network therefore visits in a sequence all the knoxels of the stored perception acts.

The denotation function ϑ describing the block C of our architecture of Figure 9.1 is implemented by setting of parameters of the energy function E to $\lambda < 1$ and $\varepsilon > 0$. The task of the block C is the recognition of input knoxel sequences representing the input perception acts. In order to accomplish this task it is necessary to consider the input term of the energy in order to make the transitions among knoxels happen as driven from the external input.

When $\lambda < 1$ the term λE_2 is not able to drive itself the state transition among the knoxels of the perception act, but when the term εE_3 is added, the contribution of both terms will make the transition happen. The neural network therefore recognizes the input perception act as it 'resonates' with one of the perception acts previously stored.

9.13 EXPERIMENTAL RESULTS

In this section, experimental results obtained by the implementation of the described architecture are presented, starting from the real image showed in Figure 9.3. Figure 9.7 shows the 3D approximation of the scene through

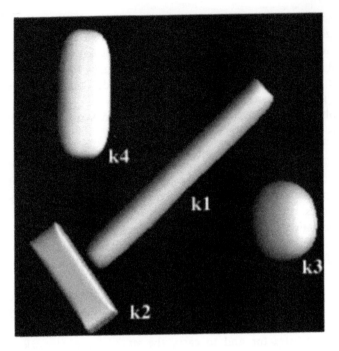

Figure 9.7 The approximation of the scene by superquadrics.

superquadrics. For the sake of clarity each superquadric has been marked by a tag.

When the described architecture is in 'reactive' modality the focus of attention searches for generic objects in the scene. In the case of the current scene, the focus of attention is directed to the hammer handle, corresponding to knoxel #k1. The knoxels related to this perception act are sent to the naming block of the architecture in order to find the corresponding linguistic constant at

Table 9.2 The assertions generated from the linguistic level describing the denomination operation of the architecture

Knoxel (#k1)
Knoxel (#k2)
Knoxel (#k3)
Knoxel (#k4)
Input_sequence (#k1)
Cylinder_shaped (#k1)

the linguistic level. In this case the knoxel #k1 has been recognized as a *cylinder_shaped* knoxel.

The assertions generated from the linguistic level describing the operation of the architecture are reported in Table 9.2.

The expectation functions suitably drive the focus of attention in order to find the relevant perception acts in the scene. The linguistic expectation function, in particular, generates hypotheses by inferences at the linguistic level. As an example with reference to the previous scene, the linguistic level hypothesizes that the cylinder shaped knoxel may be a filler for the role *hammer handle* of the concept *hammer*. Therefore the architecture hypothesizes the presence of an hammer in the scene; the linguistic expectation block generates perceptions act hypotheses for the fillers of the role part *hammer head* and for the filler of the role part *hammer handle*.

The time-delayed neural networks implementing the linguistic expectation block of the architecture generates the expected perception acts for the hammer-head filler and the hammer-handle filler. When some of these expected knoxels match some corresponding knoxels in the scene, the corresponding perception act is sent to the denomination function in order to recognize the hammer. The resulting assertions generated at the linguistic level are reported in Table 9.3.

The task of the associative expectation function in block E is to suitably drive the focus of attention in order to explore the scene by free associations among concepts.

As an example, referring to the previous scene, the concept *hammer* is associated by a Hebbian mechanism to the concepts of *ball* and *mouse*, due to a previous learning phase of the architecture. The linguistic level therefore hypothesizes the presence of these objects in the scene and the time-delayed neural network implementing the associative expectation block generate the corresponding perception acts hypotheses.

As in the linguistic modality, when some of the expected knoxels are satisfied by some corresponding knoxels in the scene, the perception act made up by the so found knoxels is sent to the denomination function. Table 9.4 shows the corresponding generated assertion by the linguistic level.

Table 9.3 The assertions generated from the linguistic level describing the linguistic expectations operation

---

Linguistic_expectation (Cylinder_shaped, Hammer)
Expected (Hammer, Hammer_head_filler)
Expected (Hammer, Hammer_handle_filler)
Satisfied_by (Hammer_head_filler, #k2)
Satisfied_by (Hammer_handle_filler, #k1)
Hammer (#k1, #k2)

Table 9.4 The generated assertions from the linguistic level describing the associative expectations operation

---

Associative_expectation (Hammer, Ball)
Associative_expectation (Hammer, Mouse)
Expected (Hammer, Ball)
Expected (Hammer, Mouse)
Satisfied_by (Ball, #k3)
Satisfied_by (Mouse, #k4)
Ball (#k3)
Mouse (#k4)

The recognition task of perception acts related to spatial relation is similar to the recognition task of objects; Table 9.5 shows the generated assertions related to spatial relations referred to the previous scene. Figure 9.8 shows the resulting perception act of the architecture after the previously described operations of the focus of attention.

Figure 9.9a shows a complex scene made up by a hammer, a cordless telephone, a wood block and a mouse. Figure 9.9b shows the superquadric reconstruction of the same scene along with the focus of attention movements through the scene exploration. It should be noted that the focus of attention follows two sequences: a sequence in which the attention is focused on the hammer, the block and the mouse, and another sequence in which the attention is focused on the body and the antenna of the telephone. The scene is therefore analysed as a concatenation of these two sequences.

The focus-of-attention mechanism allows in fact the creation of 'attentional contexts' in which an object is analysed. During the analysis of the first sequence, the telephone has been ignored, because the object does not belong to the current attentional context. The same during the second sequence: the block, the hammer and the mouse are ignored because they do not belong to the same attentional context of the telephone.

This allows the system to avoid the 'cognitive overload' problem. The system is able to find out the relevant paths, to aggregate the information, in order to generate only the linguistic descriptions 'useful' and 'interesting' for the system in the current attentional context.

Table 9.5 The generated assertions describing the spatial relations among objects in the previous scene

---

Up (Hammer, Ball)
Sided (Ball, Mouse)

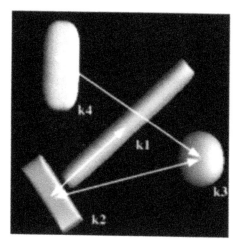

Figure 9.8 The resulting perception act related to previous scene when the focus of attention is driven by the linguistic and associative expectations.

9.14 CONCLUSIONS

Some open questions arise from the operation of the proposed architecture. One of them has already been mentioned and it is related to the lack of capability of an effective 3D scene segmentation. The segmentation into meaningful parts is currently made up by directly acting on the input images; more complicated

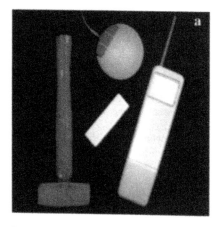
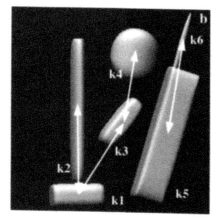

Figure 9.9 A complex scene made up by a hammer, a cordless telephone, a wood block and a mouse. a) The acquired scene. b) The superquadric reconstruction along with the focus-of-attention operation.

input scenes require most effective segmentation tools. To this aim, methods based on the application to 3D shapes of morphological tools like erosion and dilation seems to be promising.

Another open question is a truly adequate best fitting procedure, as far as the superquadric recovery is concerned. Neither common minimization techniques nor optimization procedures of the kind illustrated in the paper seem to be completely reliable, due to the local minima problems. Possible solutions are based on the direct, one-shot, estimates of the superquadric parameters by a search over a rich catalogue of prototypical superquadrics.

REFERENCES

Ardizzone, E. and Ardizzone, L. (1995) Modeling solid shape: pure and deformed superquadrics. Unpublished work.

Ardizzone, E., Gaglio, S. and Sorbello, F. (1989) Geometric and conceptual knowledge representation within a generative model of visual perception. *Journal of Intelligent and Robotic Systems* **2**, 381–409.

Ardizzone, E., La Cascia, M. and Pirrone, R. (1995) Boolean operations between superquadric-based shapes. Unpublished work.

Bajcsy, R. and Campos, M. (1992) Active and exploratory perception. *Computer Vision, Graphics and Image Processing: Image Understanding* **56**, 1, 31–40.

Barr, A.H. (1981) Superquadrics and angle-preserving transformations. *IEEE Computer Graphics and Applications* **1**, 11–23.

Biederman, I. (1985) Human image understanding: recent research and a theory. *Computer Vision, Graphics and Image Processing* **32**, 29–73.

Birnbaum, L., Brand, M. and Coooper, P. (1993) Looking for trouble: using causal semantics to direct focus of attention. *Proc. ICCV-93*, 49–56.

Bolle, R.M. and Vemuri, B.C. (1991) On three-dimensional surface reconstruction methods. *IEEE Trans. on Pattern Analysis and Machine Intelligence* **PAMI-13**, 1–13.

Brachman, R.G. and Schmolze, J.G. (1985) An overview of the KL-ONE knowledge representation system. *Cognitive Science* **9**, 171–216.

Burt, P.J. (1988) Smart sensing within a pyramid vision machine. *Proceedings of the IEEE* **76**, No. 8.

Chella, A., Frixione, M. and Gaglio, S. (1994) *A cognitive architecture for artificial vision.* CS&AI Lab Technical Report, University of Palermo.

Dickinson, S.J., Pentland, A.P. and Rosenfeld, A. (1992) 3-D Shape recovery using distributed aspect matching. *IEEE Trans. on Pattern Analysis and Machine Intelligence* **PAMI-14**, 174–198.

Farah, M.J.K., Hammond, D., Levine, R. and Calvanio, R. (1988) Visual and spatial mental imagery: dissociable systems of representation. *Cognitive Psychology* **20**, 439–462.

Gärdenfors, P. (1992) *Three levels of inductive inference.* Lund University Cognitive Studies No. 9, Technical Report LUHFDA/HFKO-5006-SE.

Glasgow, J.I. (1993) The imagery debate revisited: a computational perspective. *Computational Intelligence* **9**, 4.

Hopfield, J.J. (1982) Neural networks and physical systems with emergent collective computational abilities. *Proceedings of the National Academy of Sciences USA* **79**, 2554–2558.

Horn, B.K.P. (1986) *Robot vision*. MIT Press, Cambridge, Massachusetts.

Horn, B.K.P. and Brooks, M.J. (1989) The variational approach to shape from shading. In: *Shape from shading* (eds B.K.P. Horn and M.J. Brooks), 173–214, MIT Press, Cambridge, Massachusetts.

Johnson, Laird, P.N. (1983) *Mental models*. Harvard University Press, Cambridge, Massachusetts.

Kleinfeld, D. (1986) Sequential state generation by model neural networks. *Proceedings of the National Academy of Sciences USA*, **83**, 9469–9473.

Kosslyn, S.M. (1980) *Image and mind*. Harvard University Press, Cambridge, Massachusetts.

Marr, D. and Nishihara, K.H. (1978) Representation and recognition of the spatial organization of three-dimensional shape. *Proceedings Royal Society of London* **B200**, 269–294.

Nebel, B. (1990) *Reasoning and revision in hybrid representation systems*. LNAI 422, Springer-Verlag, Berlin, Germany.

Pentland, A.P. (1986) Perceptual organization and the representation of natural form. *Artificial Intelligence* **28**, 293–331.

Pentland, A.P. (1987) Recognition by parts. *Proc. of IEEE 1st Int. Conf. on Computer Vision*, 612–620.

Pentland, A.P. (1989) Local shading analysis. In: *Shape from shading* (eds B.K.P. Horn and M.J. Brooks), 443–487. MIT Press, Cambridge, Massachusetts.

Pentland, A.P. (1990) Linear shape from shading. *Int. Journ. of Computer Vision* **4**, 153–160.

Posner, M.I. (1980) Orienting of attention. *Quarterly Journal of Experimental Psychology* **32**, 2–25.

Putnam, L.K. and Subrahmanyam, P. (1986) Boolean operations on *n*-dimensional objects. *IEEE Computer Graphics and Applications*, 43–51.

Requicha, A.A.G. (1980) Representations for rigid solids: theory, methods and systems. *ACM Comp. Surveys* **12**, 4, 437–464.

Rimey, R.D. and Brown, C.M. (1994) Control of selective perception using Bayes nets and decision theory. *Int. Journal of Computer Vision* **12**, 2/3, 173–207.

Solina, F. and Bajcsy, R. (1990) Recovery of parametric models from range images: the case of superquadrics with global deformations. *IEEE Trans. on Pattern Analysis and Machine Intelligence* **PAMI-12**, 131–147.

Tilove, R.B. (1980) Set membership classification: A unified approach to geometric intersection problem. *IEEE Trans. Computers* **29**, 874–883.

Tsai, P.S. and Shah, M. (1992) *A simple shape from shading algorithm*. Technical Report CS-TR-92-94, Univ. of Central Florida, Orlando, Florida.

Yarbus, D.L. (1967) *Eye motion and vision*. Plenum Press, New York.

Appendix 9.A

A Brief Description of the KL-ONE System

The principal elements of the KL-ONE system descriptions (Brachman and Schmolze, 1985) are concepts; the most important type of concept is the generic concept. Potentially many individuals in any possible world can be described by a generic concept; so *object, hammer, screw*, etc. are all generic concepts each of which are descriptions that could be used to describe individuals in the world. Some of these generic concepts may be formed out of the others: e.g. a *hammer* is a *complex object*. KL-ONE separates its descriptions into two basic groups: primitive and defined. Primitive concepts are those for which necessary and sufficient conditions are not given in terms of other concepts; they act as incomplete descriptions. Defined concepts are instead derived from other concepts. In the KL-ONE descriptions the concepts are depicted with ellipses; the concept with an asterisk are primitive, while the others are defined (see e.g. Figures 9.5 and 9.6).

Several structure-forming operations are available for building concepts. They bring together one or more general concepts and a set of restrictions on the concepts. More in details, the components of a concept are its subsuming concepts and its local internal structure expressed in roles and structural descriptions. Roles describe potential relationships between instances of the concept and those of other associated concepts (e.g., its parts and properties). Structural descriptions express the interrelations among the roles; they are not used in our architecture.

Roles of a concept are taken as a set of restrictions applied to its subsuming concepts. Because a concept is defined in terms of the subsuming concepts, all the parameters restrictions must apply to the children. In order to achieve this effect, KL-ONE provides inheritance facilities: e.g., in Figure 9.5 *hammer* inherits all the component restrictions of *complex object*.

A role acts as a generalized attribute description, representing potential relationships between individuals of the type denoted by the concept and other individuals. Generic rolesets are the most important type of roles. They capture the notion that a given functional role of a concept can be played by several different entities for one individual. In KL-ONE representations rolesets are depicted by squares, connected by unnamed links to the concept of which they are components.

The structure of the rolesets is specified with value restriction (V/R) describing necessary type restrictions on rolesets fillers, and with value number restrictions, expressing cardinality information as a pair of numbers, defining a range of cardinality for sets of role-player descriptions (the nil specification means an infinite upper bound). Roleset restriction allows to describe how the roleset components of a concept can be specified in terms of rolesets belonging to the subsuming concepts. A roleset restriction does not specify a new role but it adds constraints on the filler of a role with respect to a specified concept. Roleset differentiation allows to specify the sub-roles to fill with subsets of the fillers of the roles they differentiate.

PART III

VISUAL COMMUNICATION

We have been over-enthusiastic about automatic vision tasks: unfortunately, many 'general-purpose' image-processing algorithms have failed when used in specific contexts, and no general theory for recognizing patterns has succeeded in obtaining efficient and effective means to solve real-world problems. Rather than designing *ad hoc* systems to tackle problems in specific domains, the available human knowledge from given working areas can be used, provided that a good communication exists between artificial vision programs and experienced users. In this way, a dialogue between programs and users may be usefully developed, so as to change the nature of computing into an interactive one, where a number of conditions must be satisfied in order to make it fruitful, fast and enjoyable. To this aim, the nature of images has evolved, from only being objects to be recognized into objects to be used within a communicating alphabet (iconic flavor). Recent programs consider images as descriptions recognizable by the machine, having a well-defined set of meanings to be interpreted by programs, or viewed as iconic commands to be interpreted by humans.

Chapter 10 considers illusion and its different interpretations as a starting point to describe the consequences of virtual reality on an observer. Idols and icons are introduced to describe the processes of inclusion and immersion, and the impact that different representations have on humans. Chapter 11 attempts to unify different areas of pattern recognition, image interpretation and computer graphics, by considering images not only as objects to be analysed but also as valuable communication tools. A unified language, based on the concept of visual sentence, allows the design of visual interactive environments for working with images. Chapter 12 discusses some peculiar features of pictorial databases, with emphasis on those containing sequences of images, which are relevant for multimedia applications. In such cases the retrieval process should be content based, instead of being indexed in a traditional way. Two main approaches are emphasized: symbolic indexing and iconic retrieval. Chapter 13 extends the domain of visual languages to enable the manipulation of multimedia objects using special knowledge structures, in such a way that objects themselves can both communicate efficiently, and react to events. A number of icons are then described, both theoretically and practically.

10

Illusion and Difference

10.1   THE NATURE OF ILLUSION

The historian of art Ernst Gombrich has observed:

> 'Illusion ... is hard to describe or analyse, for though we may be intellectually aware of the fact that any given experience must be an illusion, we cannot, strictly speaking, watch ourselves having an illusion'[1].

Starting with this consideration and assuming that each object and each model have a meaning and are always perceived within a given cultural system of reference, illusion could be defined as the annihilation of the perception of the difference between the object and its representation, between the model and its copy.

But, giving such a definition, the notion of difference becomes crucial, or, more precisely, as just said, illusion is given by the annihilation of the perception of difference. Such a statement assumes, as a consequence, that the perception of difference deals, at least, with that aspect of the information that does not make us identify and mistake the object for its representation, the model for its copy. In fact the notion of difference is decisive to understand the concepts of information and communication. As a consequence it is necessary to make a reference to the considerations made by Gregory Bateson, an anthropologist and a philosopher, who, refusing western dualism, takes into consideration notions such as interface or context, not as places of separation (for example between external and internal, focus and background, between

---

[1] Gombrich (1961), pp. 5–6.

ARTIFICIAL VISION
ISBN 0-12-444816-X
Copyright © 1997 Academic Press Ltd
All rights of reproduction in any form reserved

appearance and reality, mind and body, etc.), but, on the contrary, as those of information and communication[2].

If difference enables us to grasp the information, as in painting or in photography, which are not simply trompe-l'oeil, as in cinematography (except for some films or some sensational scenes[3]) or in TV, there is not any illusion, or at least this kind of illusion to which Gombrich refers to, which seems to be identified with deception. The frame makes of a picture a picture. It gives it the border of meaning and sense, that is to say that it creates a context that makes what is described in the picture peculiarly comprehensible.

Frame creates the difference between the meanings contained within the picture and the world outside. It is the same in photography, in cinematography and TV. It is true, as Gombrich says, that it is impossible that one observes oneself in the act of yielding to an illusion, but this happens when one loses the sense of difference, that is to say when one deceives oneself on the nature of what one sees, as in trompe-l'oeil, or when one identifies oneself to the point of forgetting that one is witnessing as in cinematography. In general, in any case, things are not in this way, because frame avoids turning illusion into deception, marking the limits of the context of the meanings which are within its borderlines[4]. As a consequence one, looking at a picture, a photograph, a film, a broadcast, thanks to the frame which produces difference, is warned about what one is watching and the illusory nature of the High Definition Technologies (HDT) does not turn into deception. Because of the frame the dream of the HDT of identifying the copy with the model cannot be realized. The copy, as a deceptive substitute of the model, rarely succeeds in escaping from the border-lines imposed by the frame and, as a consequence, nearly always tends to betray its nature of copy.

However, if the definition given by Gombrich about the nature of an illusion, does not seem to be pertinent as far as HDT are concerned such as prospect, photography, cinematography, TV, it is as far as Virtual Reality (VR) is concerned.

10.2 THE 'FINESTRA ALBERTIANA' AND VIRTUAL REALITY

There are at least two reasons imposing this question.

a)   The first one is about the fact that the HDT producing fixed images[5] such as perspective painting, photography and the HDT producing images in

---

[2] Bateson (1979); Bateson and Bateson (1987).

[3] I am referring to the 'subjective' scenes taken, that is to say to the scenes taken so to show the spectator what he would see if he was in the place of the protagonist. In general many scenes in horror or thriller films are in 'subjective'. Cf. Gallarini (1994), p. 26.

[4] As Svletana Alpers has observed as far as *Art and Illusion* is concerned: 'Gombrich concludes by defining a perfect representation as indistinguishable to our eyes from nature' (Alpers, 1983; p. 36).

[5] About the meaning and the role of fixed images, Freedberg (1989).

movement, such as cinematography and TV are within a principle that can be called the principle of the 'finestra albertiana', explained later one. One wonders if VR does not allow this principle based on frontality to be overcome.

b) The second fact that VR assumes the inclusion of the observer within the context of observation, so that the observer can become an actor. It is true that the previous HDT anticipated this possibility (let us think, for example, about *Las Meniñas* by Vélasquez, in painting), but with VR it has become an integrating and necessary part to the point of weakening the principle of frontality, that is to say the relationship between the position of the observer and the illusion of the three-dimensional nature of images.

The frontality and the separation of the observer from the context of observation have been two characteristics of the modern conception of the subject-object relationship and it was the Renaissance perspective painting, born in Florence thanks to Filippo Brunelleschi[6], which started the modern idea of High Definition Technology. As already studied, since the twenties', thanks to the art historian Erwin Panofsky, following Ernst Cassirer, a philosopher in symbolic forms, the 15th century invention of the pictoric perspective created by Filippo Brunelleschi, and mathematically expressed in *De pictura* by Leon Battista Alberti, gave birth to the modern idea of space with the determination of that relationship between subject and object which would then be theorized by Descartes[7].

The perspective of the 15th century, which shows the image in the picture as through a window[8], gives the modernity the three following characteristics:

a) The maximum of the reproduction of the real and of the mimetic reality (high definition technology) corresponds to the maximum of artificiality and illusion.

b) Subject and object are one in front of the other: frontality is constitutive of knowledge.

c) The world is homogenized by mathematics.

A scholar in multimediality, De Kerckhove, has observed that VR is an overcoming of the relationship subject-object based on frontality and a victory on 'the tyranny of glance'. According to De Kerckhove, we were used to seeing the world partially, always in a frontal way, at the level of our eyes, nothing above them, nothing below them, nothing under our feet, influenced by Renaissance painting and reading. The other senses which, until the apparition of perspective were dependent on environment, started to have, because of the tyranny of glance, instrumental instead of cognitive functions. First of all,

[6] Manetti (1992); Filarete (1972), vol. II, book XXIII, p. 657; Vasari (1991), see Gioseffi (1957); Edgerton (1975); Kemp (1990).
[7] E. Panofsky (1980), pp. 67–68.
[8] See Alberti (1980), pp. 36–37.

they are useful to inform the point of view before contributing to link people with each other. The apparition of the virtual also demands a transformation of the strategies of the visual control[9].

Myron Krueger, a theorist and a planner of *artificial reality*, says, regarding his strategy of research, that the aim of his work is of connecting mind and body in a new intellectual experience of the world. According to Krueger, since the invention of writing the life of the mind has been separated from the dinamicity of the body, which had to remain inactive given that only the writing and reading techniques were used. Now it is impossible to commit the body again to the research of knowledge and of the composed experience[10].

As a consequence, we are in a strategic and theoretical context, where VR wants to propose itself as an answer, and probably as a solution, to the western dualisms of subject and object, of mind and body, to frontality, to the domination of the eye over the other senses, and so as an answer to the limits of HDT, which are a heritage of the 15th century perspective painting.

But, besides any (though comprehensible) emphasis, the consequences (negative and positive ones) of the achievement and development of VR are not predictable. There is no doubt, in any case, that the idea of overcoming the principle of frontality and the attempt to eliminate the separation of the observer from the context of observation is very interesting, epistemologically and philosophically speaking, and deserves a very careful remark both from a sociological and in a psychological point of view, but also as far as relationships within the built environment are concerned.

Furthermore, evidently as the idea of overcoming the principle of frontality such as that of including the observer in the observation acquires a very important meaning in the matters of communications and learning. It could be important to remark on one hand the relationship between body, communication and learning and, on the other hand, the distinction between dualism, implying a separation between two poles, and duality, implying, on the contrary, a difference and, at the same time, a relationship between two poles (the borderline as an interface)[11].

Theoretically speaking, a tendency of HDT is to create effects of illusion, that is to say reaching such a perfection in building a copy so that the copy itself becomes the model and substitutes it thanks to its greater power and perfection. But this tendency of HDT, whose hypothetic perfection could exist with the achievement of a perfect identity, seems to be, in any case, a psychological and literary nightmare of man, who is substituted by his double. Between this tendency of HDT and the nightmare of man is that whole of identity and diversity which, in different ways, is at the head of our symbolic production, communication and learning: our double (*homo duplex* as Buffon, Baudelaire and Durkheim say), whose image is shaken each time we are in front of an

[9] D. De Kerckove, *Psicologia postmoderna nell'arte della realtà virtuale*, in Belotti (1993) p. 62.

[10] M. Krueger, *Realtà Artificiale. Arte versus azione*, in Belotti (1993), p. 155.

[11] About this, Bateson (1979) and Bateson and Bateson (1987).

epistemological and cognitive uneasiness. According to John Raltson Saul, the author of *Voltaire's bastards*, the electronic perfection of image has been the final step of the research for idolatry taken by western man, which started with the integration of the Roman pagan and rational foundations in the Christian Church, then by the important step of completing the perfection of the static image, thanks to Raffaello who painted the Athens principles for a Renaissance Pope, until reaching its epilogue today. The fear consuming the human soul is a reflection of this conclusion. It is as if our image and we were going around an eternal circle, looking at ourselves the one in the other suspectedly and intensively[12].

10.3 ABOUT THE RELATIONSHIP BETWEEN IDOL AND ICON

Suzanne Said, has observed, also following the consideration made by the philosopher Jean-Luc Marion, that the distinction between idol and icon in the ancient Greece is characterized by an opposition which sees in the idol the copy of the sensitive appearance and in the icon a transposition of the essence. What Suzanne Said wants to point out is the fact that idol, differently from icon, would mainly deal with the sphere of the visible. Its main characteristic is the deceiving resemblance: the idol tends to mingle with what it represents and, as a consequence, to substitute it. According to Said, in origin the archaic idol is not the product of the human industry, but after that the resemblance of the image has become the product of a human art, it has not been admired anymore as a miracle, but the phenomenon itself of resemblance started to be analysed and the illusion of the trompe-l'oeil[13] started to be denounced. All the contrary happened for icon, which opposes itself to idol character, due to its role of symbol and not due to that of simulacrum. The icon does not want to substitute the model and create a deception. The idol pretends to be its model and tries to mingle with it. The icon recognizes itself as different from the idol and only wants a relative identity[14].

The ideological implications are evident in this work by Suzanne Said, which has its roots in the difficult discussion about images concerning the great monotheistic religions: the opposition idol/icon tends to save conceptually at least an aspect of the role of images (icons) from their tendency to be deceiving substitutes of the model they should represent.

Things, however, do not seem solved at all. Jean-Pierre Vernant has pointed out that, at least in the archaic period, between idol and icon there is not any opposition. Idol and icon, in fact, are not contemporary, since *Eikòn* is not used

---

[12] Ralston Saul (1992).
[13] Said (1987), p. 314.
[14] *ibidem*, pp. 329–330.

before the 5th century[15]. Before Plato the word *mimeisthai*, associated with the tragedy, deals with the relationship between the imitator and the spectator who watches. Plato remarks the relationship between the imitator and the imitated thing[16]. From this point of view, Plato, opposing more deeply the appearing to the being, separating the one from the other instead of associating them in different balances as it had happened previously, makes the image independent, with its own existence[17]. Before Plato, as shown[18], a precise distinction between appearing and being is not given. But things are in this way, Vernant concludes, as far as the archaic period is concerned, it is not possible to think about a precise distinction idol/icon which, as a consequence, cannot be used as a reference for a survey on the statute of image, its functions, its nature[19].

According to Vernant, the *eidolon* in the Archaic Greece is not an image in the same sense we mean it, who are accustomed to the opposition between appearance and essence, but a real presence which shows itself, at the same time as an absence, which, showing itself effectively here, reveals itself as another itself, belonging to another world. In the Archaic Greece idol is more a double than an image in the sense we mean[20].

10.4 DUALISM BETWEEN APPEARANCE AND REALITY

What Vernant wants to say here is the fact that assimilation of the notion of double to the notion of image as a possible deceiving substitute of the model occurs when, as it has been observed, the platonic distinction between appearing and being has come to pass.

Before Plato there was not, in Greek thought, an idea of dualistic opposition between appearance and reality, on whose conceptual basis the matter of a representative which becomes a deceiving substitute of the represented model can occur.

It is starting with the opposition between appearance/reality that Plato, in Respublica[21] and in Sophist[22], can condemn the representation by images, such as the pictorial one. But a condemnation of the kind makes a sense (beyond the opinion about Plato's position towards visual art[23]) only if an ability of illusion and deception in painting is assumed. That is to say that only if painting reveals

---

[15] Vernant (1990), p. 231.
[16] Vernant (1979), pp. 106–108.
[17] *ibidem*, p. 31.
[18] A. Rivier (1956).
[19] Vernant (1979), p. 233.
[20] J.-P. Vernant (1979), p. 111. Ref. to Anticleia, Ulysses' mother: *Odyssey* **11**, 153–222; and to Patroclo: *Iliad*, **23**, 65–108.
[21] Plato, *Respublica*, X, 595 and following.
[22] Plato, *Sophist*, 266a–268d.
[23] About this, Keuls (1978).

itself as an art capable to produce deceiving substitutes of the represented model within a context in which the opposition between appearance and reality, it reveals itself as an *a priori*. In the popular example of the bed and the carpenter, book X of Respublica, Plato, in order to consider the art of the painter who paints a bed as an art producing illusion, must assume the fact that an observer may mistake the bed of the painter for the bed of the carpenter. This possibility of deception can be given only if the painter is technically able to give the illusion of the third dimension.

In any case the opposition illusion/reality corresponds to the opposition appearance/reality[24]. Within these contexts, the technical ability of the producer of deceiving substitutes lies in diminishing the perception of the difference between the two poles of attraction as much as possible, or, in other words, in reaching the highest degree of resemblance between the copy and the model.

But it is necessary to specify that a reality independent from the system to which it refers to, does not exist. As far as painting and the idea of realism as a representation resembling the object are concerned, the philosopher Nelson Goodman has observed:

> 'realistic representation ... depends not upon imitation or illusion or information but upon inculcation. Almost any picture may represent almost anything; that is, given picture and object there is usually a system of representation, a plan of correlation, under which the picture represents the object. How correct the picture is under that system depends upon how accurate is the information about the object that is obtained by reading the picture according to that system. But how literal or realistic the picture is depends upon how standard the system is. If representation is a matter of choice and correctness a matter of information, realism is a matter of habit. Our addiction, in the face of overwhelming counter-evidence, to thinking of resemblance as the measure of realism is easily understood in these terms. Representational customs, which govern realism, also tend to generate resemblance. That a picture looks the way nature is usually painted. Again, what will deceive me into supposing that an object of a given kind is before me depends upon what I have noticed about such objects, and this in turn is affected by the way I am used to seeing them depicted. Resemblance and deceptiveness, far from being constant and independent sources and criteria of representational practice are in some degree products of it'[25].

10.5 BETWEEN INCLUSION AND IMMERSION

Only abstractly and theoretically speaking, it might be possible to think that if the aim of HDT is that of producing a copy as exact as possible to the model to mingle with it, deceiving in this way the observer, it is in contrast with a basic

---

[24] The mith of the cave in VII book, *Respublica*, 514a–519b. About this, Voegelin (1966); Heidegger (1976); Gaiser (1985); Blumenberg (1989); Arendt (1990).
[25] Goodman (1968), pp. 38–39.

cognitive element of the human mind, that is to say the idea of difference, which makes us perceive both the context within which information and communication are produced and the meaning of information and communication, which is not possible to be obtained without a context and a difference.

We are obviously following a schematic line of conduct which, in any case, can make a sense only if we take into consideration the fact the VR is producing, both from a cognitive point of view and one concerning social forms of life, new and interesting (but also worrying) problems within the relationship between imaginary and reality with VR, in fact we are, not only in front of a technology wanting to build a copy exactly alike the model, but, as it has been already said, the ambition is to build a copy capable, in a certain sense, to be independent from the model. In other words, VR does not just simulate[26] reality, but also produces an independent world which produces on its turn an imaginary and a reality[27].

In any case, two lines of consideration are possible.

1) The role of the observer in VR and its possible modifications towards other HDT.

It has been previously said that VR seems to confute two principles which have characterized the role of the observer in a modern way: frontality and the detachment from the observer. In a certain sense, these two principles, supported by the metaphor of the 'finestra albertiana' are based on the idea that the reproduction of the real can occur simulating the tridimensionality in a bidimensional plane. But with VR this seems to be overcome[28].

And, in any case, theoretical problems, which deserve a project of research, are present. One of the recurring metaphors concerning VR is that of immersion which has the intention of indicating the process of inclusion of the observer[29]. This would place VR outside the two previously mentioned principles, because if frontality and the detachment of the observer are within the metaphor of the 'finestra albertiana', this metaphor does not imply an immersion but excludes it. In fact, the 'finestra albertiana' works correctly only if it is perceived as it is[30], that is to say only if the observer still succeeds in distinguishing copy and model, knowing that the 'finestra' is the borderline between inside and outside. The 'finestra albertiana' assumes the distinction between inside and outside, whose borderlines would appear less marked in the process of the inclusion of the observer as it is produced in VR.

To all this it is necessary to add another element. The HDT, referring to the metaphor of the 'finestra albertiana', have a low level of interactivity. Tele-viewers, for example, have no possibilities to interact with what they see

---

[26] About the concept of simulation, Bettetini and Colombo (1993).

[27] About this see also Cardoz (1994).

[28] S. Gasparini, *La costruzione dei mondi nella Realtà Virtuale*, in Bettetini and Colombo (1993), p. 77.

[29] *ibidem.*

[30] About this, Kubovy (1986).

(phonecalls and surveys are caricatures of interaction based on reciprocity). VR seems, instead, to promise a high level of interactivity. The matter deals with the metaphor of immersion and the concept of inclusion. If, in fact, in an interactive point of view, in VR the effect of the simulated immersion of the observer seems to be dominant, and the observer can have, as a psychological and epistemological consequence, the oblivion of his double, in Artificial Reality (AR) by Myron Krueger, on the contrary, the process of inclusion of the observer does not assume immersion at all, but the possibility for the observer to see himself acting[31]. From this point of view, the impression is that VR and AR, in a certain sense, as far as the role of the observer and the psychological and epistemological conditions of interactivity with the machine are concerned, are opposite.

A survey, both historiographical and epistemological on the change of the role of the observer, produced by VR in comparison with other HDT, seems to be necessary, in particular within a context of research and application which assumes, from a technological point of view, multi-mediality. It would be interesting to try to understand this problem and its possible consequences in a situation of interactivity man-machine inside a scene where several HDT, usable for the same aim, are together and within a tendency of the developments which already propose, as in the case of the comparison between VR and AR, conceptual problems which could become soon problems of usability.

2) Images and learning.

In one of his *Lezioni americane*, precisely in that one concerning Esattezza, Italo Calvino speaks about an epidemic plague which has affected the language, an epidemic which reduces language to generic formulas, to the dilution of meanings, to the loss of the cognitive force. And he adds

'Vorrei aggiungere che non è soltanto il linguaggio che mi sembra colpito da queste cose. Anche le immagini, per esempio. Viviamo sotto una pioggia ininterrotta di immagini; i più potenti media non fanno che trasformare il mondo in immagini e moltiplicarlo attraverso una fantasmagoria di giochi di specchi: immagini che in gran parte sono prive della necessità interna che dovrebbe caratterizzare ogni immagine, come forma e come significato, come forza d'imporsi all'attenzione, come ricchezza di significati possibili. Gran parte di questa nuvola d'immagini si dissolve immediatamente come i sogni che non lasciano traccia nella memoria; ma non si dissolve una sensazione di estraneità e di disagio'[32].

The uneasiness Calvino feels is because of 'la perdita di forma' (the loss of form). This is a paradoxical situation. During the age of the great development of the high definition technologies, of precision and exactness, Calvino denounces a loss of what is essential for each learning and each communication: the form. In fact, as known, to inform means to give a form. A loss of a form is a loss of information. It almost seems that the more the means to have information about the world and the environment we live in are developed, the

---

[31] S. Gasparini, *La costruzione dei mondi possibili nella Realtà Virtuale*, in Bettetini and Colombo (1993), p. 83. But see Krueger (1991).
[32] Calvino (1988), pp. 58–59.

more the aim for which these means have been projected and realized is dissolved. Things are not necessarily in this way. However, to consider such a possibility can be important for a research about the relationship between images and learning. It would be naive and wrong to think that the coming of a new technology, surely full of new possibilities, can solve problems concerning in particular the social life. And however we are in front of new worries concerning the relationship between a medium, such as TV, and social life. Many people think that the power of a medium, such as TV, whose interactive level is almost absent, has created, in a democracy, a decrease in the taking part of the observer, marking the distance between being a spectator and being an actor. 'The plague of the language' and 'the plague of images' depend perhaps on the fact that, in a certain sense, the decisive element of communication, that is to say the centrality of the relationships among people as a decisive element for meanings and their enrichment seems to lose strength. Learning always deals with a social-symbolic relationship (starting with the relationship between mother and son). It is impossible to learn a world or an environment outside social-symbolic contexts. In everyday life, now, a medium such as TV is getting more and more substitutive of human and social relationships and, as a consequence, images do not reveal themselves as an increase in knowing possibilities but become substitutive of the language. Language gets poorer and poorer in the inevitable process of simplification which is at the head of the very short time of the advertisements of goods.

Now, our form of social life seems to go towards a struggle between language and image, and for this reason it is necessary to wonder if VR, in its several modalities, can mark even more the process of substitution of the social-symbolic relationship or if it can, on the contrary, give them more strength. It is true, in fact, that VR assumes interactivity and so a reciprocity of relationships that TV hasn't got and cannot give, but this does not assure a positive answer to the question. The fact that the notion of the inclusion of the observer can turn both into an immersion and into a seeing oneself acting presents some epistemological questions about the matter of learning. But, apart from this, the problem is the fact that the effects of simulation given by VR can either enrich the knowledge of reality (or more precisely the knowledge of our relationship with reality) or can make it poorer or even substitute it. But the answer cannot depend on the analysis of the relationship between technology and individual learning, but on reflection about the complexity of the forms of communication, within which that one of the relationship between technology and individual learning is only a particular case.

REFERENCES

Alberti, L.B. (1980) *De pictura*, ed. by C. Grayson (original ed.: 1435), Laterza, Bari, Italy.

Alpers, S. (1983) Interpretation without representation. *Representations*, **I**, 1.

Arendt, H. (1990) Philosophy and politics. *Social Research*, **1**.

Bateson, G. (1979) *Mind and nature: a necessary unity*. Dutton, New York.

Bateson, M.C. and Bateson, G. (1987) *Angels fear. Toward an epistemology of the sacred*. Macmillan, New York.

Belotti, G. (ed.) (1993) *Del virtuale*. Il Rostro, Milano, Italy.

Bettetini, G. (1991) *La simulazione visiva. Inganno, finzione, poesia, computer graphics*. Bompiani, Milano, Italy.

Bettetini, G. and Colombo, F. (1993) *Le nuove tecnologie della communicazione*. Bompiani, Milano, Italy.

Blumenberg, H. (1989) *Höhlenausgänge*. Suhrkamp, Frankfurt a/M, Germany.

Cadoz, C. (1994) *Les réalités virtuelles*. Flammarion, Paris, France.

Calvino, I. (1988) *Lezioni americane. Sei proposte per il prossimo millennio*. Einaudi, Torino, Italy.

Edgerton, S. (1975) *The renaissance rediscovery of linear perspective*. Harper and Row, New York.

Filarete, A.A. (1972) *Trattato di architettura*, 2 vol. edited by A.M. Finoli e L. Grassi (original ed. 1461–1464). Il Polifilo, Milano, Italy.

Freedberg, D. (1989) *The power of images*. University of Chicago, Illinois.

Gallarini, S. (1994) *La realtà virtuale*. Xenia, Milano, Italy.

Gaiser, K. (1985) *Il paragone della caverna*. Bibliopolis, Napoli, Italy.

Ginzburg, C. (1991) Représentation: le mot, l'idée, la chose. *Annales ESC* **6**, 1219–1234.

Gioseffi, D. (1957) *Perspectiva artificialis*. Università di Trieste, Italy.

Gombrich, E.H. (1961). *Art and illusion*. Bollingen Series XXXV-5, Pantheon Books, pp. 5–6, New York.

Goodman, N. (1968) *Languages of art*. The Bobbs-Merrill Company, Indianapolis.

Goodman, N. (1978) *Ways of worldmaking*. Hackett, Indianapolis.

Hall, E.T. (1966) *The hidden dimension*. Doubleday, New York.

Heidegger, M. (1976) *Platons lehre von der wahrheit*. In Id., *Wegmarken*, hgs. von F.-W. von Hermann, *Gesamtausgabe*, IX, Klostermann, Frankfurt a/M, Germany.

Kemp, M. (1990) *The science of art. Optical themes in western art from Brunelleschi to Seurat*. Yale University Press, New Haven.

Keuls, E.C. (1978) *Plato and Green painting*. Brill, Leiden, The Netherlands.

Krueger, M.W. (1991) *Artificial reality II*. Addison-Wesley, New York.

Kubovy, M. (1986) *The psychology of perspective and renaissance art*. Cambridge University Press, Cambridge, UK.

Levialdi, S. and Bernardelli, C. (eds) (1994) *Representation. Relationship between language and image*.

Manetti, A. (1992) *Vita di Filippo Brunelleschi*, edited by C. Perrone (original ed. betw. 1482 and 1489?), Salerno Pub., Roma, Italy.

Marion, J.-L. (1991) L'idole et l'icône. In Id., *Dieu sans l'etre*, PUF, Paris, France.

Panofsky, E. (1980) *La prospettiva come 'forma simbolica' e Altri Scritti*, Feltrinelli, Milano (original ed.: *Die perspektive als 'symbolische form'*, Teubner, Leipzig-Berlin 1927).

Parronchi, A. (1958) Le due tavole prospettiche di Filippo Brunelleschi, *Paragone*, n. 107.

Ralston Saul, J. (1992) *Voltaire's bastards. The dictatorship of reason in the west*. MacMillan, New York.

Rivier, A. (1956) Remarques sur les fragments 34 et 35 de Xénophane, *Revue de Philologie*, **XXX**, Italy.

Saïd, S. (1987) Deux noms de l'image en Grec ancien: idole et icône, *Comptes Rendus de l'Académie des Inscriptions & Belles Lettres* (avril–juin).

Vasari, G. (1991) *Le Vite de' Più Eccellenti Architetti, Pittori, et Scultori Italiani da Cimabue Insino a' Tempi Nostri*, ed. by L. Bellosi e A. Rossi, Einaudi, Torino (original ed.: 1550).

Vernant, J.-P. (1979) *Naissance d'images*. In Id., *Religions, histoires, raisons*, Maspero, Paris, France.

Vernant, J.-P. (1990) Figuration et image, *Metis*, **5**.

Voegelin, E. (1966) *Plato*. Louisiana Univ. Press, Baton Rouge, Louisiana.

11

Computing with/on Images

11.1 INTRODUCTION

In the recent years, the problem of reasoning with/on images has been addressed from several points of view. In Reiter and Mackworth (1989/90), for example, the problem of image interpretation and depiction is considered, in Wang and Lee (1993) that of visual reasoning, in Chang (1986), Wittenburg et al. (1990) and Golin (1991) definitions are suggested for visual languages and related compilers. Chang (1995) also proposed active indexes as a tool for efficient image manipulation, i.e. storage, retrieval and reasoning. However the execution of real tasks, such as deriving a diagnosis from a biopsy or designing a mechanical part with a CAD system, requires computing with/on images, i.e., the ability to execute image analysis, synthesis and manipulation within the same session.

To perform the above activities we propose a common framework that formalizes the image, the description domains and the mapping between the two: such domains are described as formal languages. This proposal stems from the study of the user-machine main channel of communication – the visual one – and allows the specification, design and implementation of systems to execute the different activities on or with images. In a typical interactive session, a

ARTIFICIAL VISION
ISBN 0-12-444816-X

Copyright © 1997 Academic Press Ltd
All rights of reproduction in any form reserved

person (from here on named 'user') looks at a display, interprets the image appearing on it to understand the state of his/her work, decides which action to perform next and interacts with the machine to have the action performed.

On the user side, the signal sent to the machine may be as simple as a button press on a certain zone of the screen or as complex as drawing a new structure on the screen. See Figure 11.1 for a typical screen layout containing the digital image of a biopsy with which a physician is interacting. For a detailed comment see Section 11.3. On the machine side, the program manages the screen bitmap: the image on the screen materializes this bitmap as organized by the program which is in control of the machine; the messages generated by the user interacting through the screen are captured exploiting the one-to-one correspondence between the bitmap and the screen elements to be interpreted by the program. In the sequel we will refer to any kind of bi-directional interaction as a user-program one.

The screen becomes the communication *surface* between the user and the program. This surface assumes the role of a bi-directional channel in the transmission of a message – the image on the screen – between user and program (Tondl, 1981).

Once captured by the system, however, the image is interpreted by the programs which control the machine following a well defined model of the

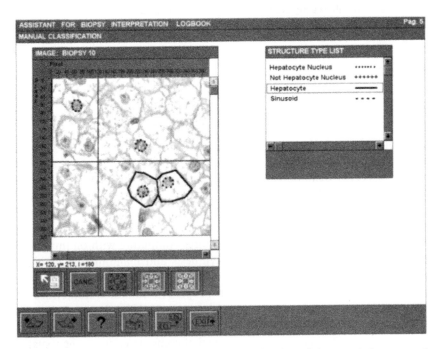

Figure 11.1 Screen display during interactive classification of characteristic patterns in a liver biopsy.

activity to be performed. This model derives from the programmer's understanding of the activity to be performed and of the user's expectations, goals and
habits – the so-called 'user model' in Kobsa and Wahlster (1989).

Visual languages will be here introduced exploring the nature of the visual
messages in the user-program communication. The way in which the program
associates a computational meaning to an image and, conversely, the way in
which it generates a pictorial description of a computation on the screen
are examined. A definition of a visual sentence as an interpreted image is
then proposed: visual languages are viewed as sets of visual sentences in a
bi-directional user-program dialogue.

From this perspective, the study of visual languages appears as the
extreme evolution of the linguistic approach to pattern recognition as it
was initially introduced by Narashiman (1969) and Shaw (1969) to derive
image descriptions which are meaningful in some knowledge domain (Fu and
Tsai, 1980).

In the last ten years, the study of visual languages has evolved into a separate
discipline to meet the new requirements created by the diffusion of end-user
computing (Brancheau and Brown, 1993) and by the integration of information
from different sources as, for example, in the manipulation of documents for the
design, management and manufacturing of industrial objects. Visual languages
intend to solve, or overcome, these problems by formalizing the languages
developed and exploited by users to express and communicate problems and
solutions in their own working environments. In such languages text, images
and diagrams are combined. Our approach to the definition of visual languages
is general and can be used to derive parameters for the analysis of existing visual
systems. Five such systems are examined showing how the proposed definition
provides a basis on which computing with/on images activities can be described.

11.2 TOWARD VISUAL LANGUAGES

Informally, a visual language is a set of images associated with their computational meaning. To make this concept precise, we first introduce the definition of
digital image as a bidimensional string. Next, the sets of images occurring in a
user-program communication are characterized as languages.

11.2.1 Pictorial languages

Definition 1 A *digital image i* is a bidimensional string

$$i: \{1, \ldots, r_{\max}\} \times \{1, \ldots, c_{\max}\} \rightarrow P$$

where $r_{\max}$ and $c_{\max}$ are two integers and P is a generic alphabet.

Definition 2   Each element of a digital image is called a *pixel*.

A pixel is described by the triple (r, c, p) where $r \in \{1, \ldots, r_{max}\}$ is the row, $c \in \{1, \ldots, c_{max}\}$ is the column and $p \in P$.

Definition 3   A *structure* is a subset of a digital image.

If the structure is rectangular, it is a 2D string itself, a substring of the original digital image and a digital image on its own.

Definition 4   A *pictorial language PL* is a subset of the set I of the possible digital images:

$$PL \subseteq I = \{i \mid \exists r, c \in \mathbb{N}, i: \{1, \ldots, r\} \times \{1, \ldots, c\} \rightarrow P\}.$$

The definition of image as a bidimensional (2D) string highlights that its elements can be ordered in several ways. On the contrary, a one-dimensional (1D) string s (defined as $s: \{1, \ldots, n\} \rightarrow V$) exploits the linear ordering only, and the consequent definition of concatenation. Linear ordering is not privileged in the analysis of 2D strings, where it is instead important to state which one of the possible orderings is to be exploited (Golin, 1991) and hence which definition of concatenation is to be adopted (Rosenfeld and Siromoney, 1993).

Several types of digital image (from now on simply image) can be defined depending on the algebraic structure which is imposed on the defining alphabet. For example:

1)   *Gray level images* (GLI) are defined on a segment of the integers: $P = \{0, \ldots, q\}$ with $q = 2^h - 1$, h being an integer. GLI can be operated on by numeric or weaker operators.
2)   *Black and white images* (BWI) are defined on $P = \{0, 1\}$. BWI can be operated on by Boolean or weaker operators.
3)   *Segmented images* (SI) are defined on a set of arbitrary labels $P = \{l_1, \ldots, l_k\}$ on which no numerical structure is imposed. SI can be operated on by nominal and order scale operators.

Note that in SI, each label l_i is in general associated with a meaning. The segmented image appears as partitioned into (possibly non-connected) sets, each set being identified by a label, which denotes its meaning.

11.2.2   Characteristic patterns

The meaning of an image may be coded as a *description* constituted by vectors of values, each value being assumed by an attribute. Values are traditionally called *features* (Haralick and Shapiro, 1991). As an example of image description we may refer to Figure 11.1. Here the window IMAGE: BIOPSY 10 shows a gray level histological image in which four nuclei and two cells have been outlined. A description of this image is $\langle 10: 4, 2 \rangle$ where 10 is the image identifier, 4 is the value of a first attribute, the CURRENT # OF HEPATOCYTE NUCLEI and 2 the value of a second attribute, the Current # of Classified Hepatocytes. Structures, being images in turn, can be similarly described. A hepatologist can, for example, describe the rightmost hepatocyte nucleus in Figure 11.1 by $\langle 3: (280, 246), 709, 99 \rangle$ where the first number is the nucleus identifier, the next two numbers are the (X, Y) Coordinates of the leftmost pixel of the structure, the fourth is its Area and the fifth is its Perimeter.

In syntactic pattern recognition, this definition has been extended to admit not only measurements in the rational scale, but also in scales which admit fewer operators, such as the nominal or the ordinal ones (Siegel, 1956). Hence the description of a hepatocyte, whose sub-structures are the fourth classified nucleus and the first classified membrane, can be:

$$\langle 7: (210, 230), 4473, 483, \text{normal}, (\text{'nucleus4','membrane1'}) \rangle$$

where 'normal' denotes the nominal value of the attribute State and ('nucleus4','membrane1') the two nominal components of the attribute Characteristic Patterns. These nominal components identify two structures in the image that the hepatologist considers important for understanding the image meaning.

In general, the description of an image requires the recognition of those structures that an observer considers necessary to understand and communicate its meaning. These structures are named 'characteristic patterns' (or 'form features' (Cugini *et al.*, 1988)).

An observer identifies the characteristic patterns of an image on the basis of their functional meaning, which derives from their geometrical, topological, chromatic and morphological properties.

The users identify characteristic patterns of similar meaning through a common name.

Definition 5 A *characteristic pattern* (or *form feature*) is a structure identified by a name.

The set of characteristic patterns defining the meaning of an image depends on the context and the goals to be achieved on or through the image. If the interpretation context changes, the set of characteristic patterns recognized in an image may also change. An example is shown in Figure 11.2 where different interpretations of a remote sensed image are singled out.

11.2.3 Description languages

An image can be described by enumerating the characteristic patterns an observer extracts from it. More precisely, an observer describes an image both by a name – as a whole characteristic pattern – and a set of features.

The name resumes the observer's overall interpretation of the image and the features are the values of the attributes the observer decides as relevant to

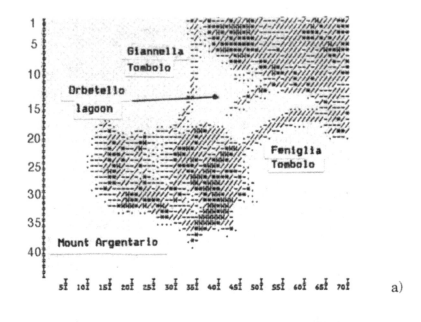

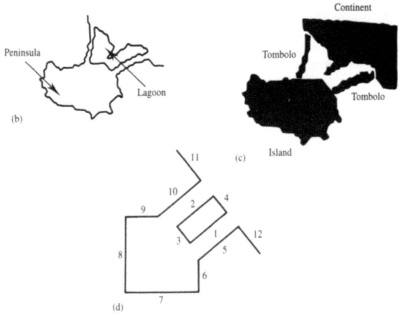

Figure 11.2 A remotely-sensed image of the Argentario promontory a); its geographic interpretation b); its geological interpretation c); a sketch outlining substructures ambiguously labelled by the two interpretations d).

describe the image. Some of these features in turn may be names of characteristic patterns. Again, the observer can describe such a characteristic pattern
by a vector containing its name and the values of its features. More precisely: let
T be the set of names of characteristic patterns and $\forall t \in T \exists A_t = \{a_1, \ldots, a_m\}$
and $\forall a_i \in A_t \exists P_{ai} = \{p_1, \ldots, p_q\}$ where A_t is the set of the attributes of the
characteristic patterns denoted by the name t and P_{ai} is the domain of the
attribute a_i (Fu and Tsai, 1980).

Definition 6 The description of a characteristic pattern is the $(m + 1)$-tuple
constituted by a name $t \in T$ and a m-tuple of properties:

$$p_{ai}, 1 \leqslant i \leqslant m \text{ and } p_{ai} \in P_{ai}$$

This description has the structure of an attributed symbol (Knuth, 1968).
Indeed: let V be an alphabet and $\forall v \in V \exists A_v = \{a_1, \ldots, a_m\}$ and $\forall a_i \in
A_v \exists P_{ai} = \{p_1, \ldots, p_q\}$, where A_v is the set of attributes of the symbol v and
P_{ai} is the domain of the attribute a_i.

Definition 7 An *attributed symbol* is the $(m + 1)$-tuple constituted by a symbol
$v \in V$ and m elements $p_{ai} \in P_{ai}, 1 \leqslant i \leqslant m$.
  We call $W = V \times Pa_1 \times \cdots \times Pa_m$ the alphabet of attributed symbols.

Definition 8 A *string sa* of attributed symbols on W is a function

$$sa: \{1, \ldots, n\} \rightarrow W.$$

Definition 9 A *description* of an image is a string of attributed symbols

$$d: \{1, \ldots, n\} \rightarrow W = T \times P_{a1} \times \cdots \times P_{am}$$

where each attributed symbol describes a structure in the image.

Definition 10 A *description language DL* is a subset of the set D of the possible
descriptions:

$$DL \subseteq D = \{d \mid \exists g \in \mathbb{N}, d: \{1, \ldots, g\} \rightarrow W\}.$$

11.2.4 A generalization of the concept of description

A description of an image can refer not only to static properties of the structure,
but also to their possible evolution in time, i.e. dynamic properties. The
specification of such evolution can be a program; in this case the description
becomes executable. For example, in an object-oriented approach each attributed symbol defines an object, in which the attributes play the role of the state
variables of the objects and the names in T play the role of names of methods.
More generally, if $t \in T$ denotes a type of characteristic pattern (say *tombolo* in
Figure 11.2a and d) in the description of an image, it may be interpreted as the
name of the class of objects associated with that type of characteristic pattern.
  This generalization imposes some constraints on the set of attributes. A full

description of the object must include:

a) state variables describing the mutual spatial relations of the structures in the image;
b) names of methods which define the executable activities associated with the described characteristic pattern.

11.3 VISUAL SENTENCES

Computing with/on images requires the integrated management of images and descriptions. In the different stages of the computation, the program can be required to describe or synthesize an image as well as to retrieve the description associated with a characteristic pattern or conversely to present the characteristic pattern associated with a given description. To perform these activities it is necessary to specify the relations among characteristic patterns in the image and the attributed symbols in the description. Figure 11.3 exemplifies the relations between the characteristic patterns of a liver biopsy and the corresponding attributed symbols in a description. The description consists of a string of seven attributed symbols. Let $nu \in T$ denote cell nuclei. The two attributed symbols with name nu describe the two nuclei and are characterized by five features: 1) numeric identifier, 2) position identifier (x, y), 3) area, 4) perimeter, 5) diagnostic value.

Characteristic patterns and attributed symbols are related in two ways: first, each characteristic pattern in the original image must be associated with an interpretation, so that a user selecting the characteristic pattern on the screen may reach such interpretation. Second, an inverse relation must be established associating at least a characteristic pattern with each attributed symbol (Chang et al., 1987). Note that the image i is itself a characteristic pattern (a 2D string) described by an attributed symbol in the description d (e.g. the symbol bp in Figure 11.3). The characteristic patterns of i are described by the other attributed symbols in d.

Hence, a visual sentence vs can be defined:

Definition 11   A *visual sentence* is a triple $vs = \langle i, d, \langle int, mat \rangle \rangle$ where i is an image, d a description and $\langle int, mat \rangle$ a pair of relations between substrings of i and d.

11.3.1 An insight on the relation between images and descriptions

Note that $\langle int, mat \rangle$ must be defined taking into account that they may relate a substring to a finite set of substrings. For example, a pixel may belong to several

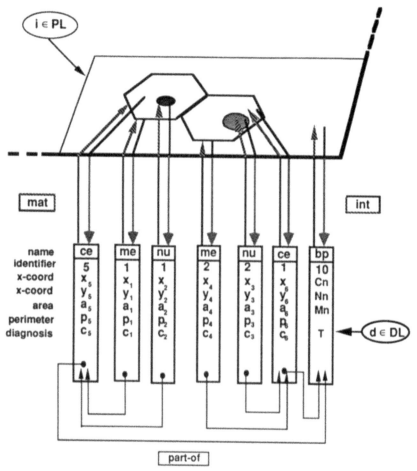

$$\text{vs} = < \text{i, d, } <\text{int, mat} >>$$

ce stands for cell, nu for nucleus, me for membrane,
a for area, p for perimeter, bp for biopsy

Figure 11.3 A schematization of the visual sentence components.

characteristic patterns (and hence be related to several attributed symbols) and, conversely, an attributed symbol may correspond to several characteristic patterns in an image. In the case illustrated in Figure 11.3, for example, pixels belonging to a nucleus also belong to a cell, due to the fact that a nucleus *nu* is part of a cell.

In a similar way as happens in Figure 11.4, a description is represented on the screen both by a facsimile, iconically and through a polygonal approximation of component characteristic patterns.

The relation *int* $\subseteq PL \times DL$ is called the *interpretation* of the image and

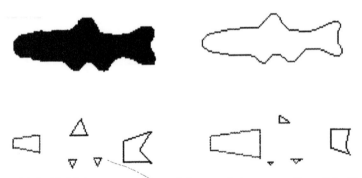

Figure 11.4 Multiple representation of characteristic patterns for a fish silhouette.

associates each pixel belonging to a characteristic pattern with a set of attributed symbols. On the other hand, *mat* $\subseteq$ *DL* $\times$ *PL* is called the *materialization* of the description and associates an attributed symbol with a set of characteristic patterns (Chang *et al.*, 1987). The structure of $\langle int, mat \rangle$ becomes more complex whenever an image *i* in *PL* is subject to different interpretations. In these cases a same set of pixels is classified as being part of different characteristic patterns. Thus, several descriptions in *DL* are associated with it and several *vss* can be defined from *i*. Analogously, the same description can be materialized into different characteristic patterns. The two phenomena are illustrated in Figures 11.2 and 11.4. For example, in Figure 11.2c the pixels in the segment 2 can be interpreted as a part of *tombolo* or as a part of the *lagoon*. In Figure 11.4, on the contrary, the same string describing the overall interpretation of the fish can be materialized in three different ways, namely by labelling each pixel in the original contour with a colour resuming the meaning of the structure to which the pixel belongs; by producing a polygonal approximation of the original contour; or by an iconic representation associating arbitrary shapes with each kind of structure. Multiple interpretation and multiple representations can then be described by defining set-valued functions. However, for our present purposes, we may treat *int* and *mat* as functions and take the liberty of overlooking their structures.

Allowing the description *d* to represent a program *P* imposes a more rigid form on the two mappings *int* and *mat*. Characteristic patterns in the image appearing on the screen are connected by tools implementing *int* to blocks in the program which drives them and, vice versa, blocks in the program *P* are connected to characteristic patterns in the image by tools implementing *mat*. In a typical user-program interaction, a user points and selects a zone of the screen, thus indicating a characteristic pattern – for example, even a button associated with a program block *m*.

The program *P* receives the coordinates of the selected pixel, identifies the computational meaning of the message sent by the user – say, to activate the program block *m* associated with the button – and then checks whether some given conditions hold and, if this is the case, activates *m*. A result of this computation can be a modification of the image or the creation of a new one.

For example, in Figure 11.1 a histologist is exploiting ABI – the automatic Assistant for Biopsy Interpretation (Mussio *et al.*, 1991a) – to interactively interpret the digitized biopsy BIOPSY 10. The pictorial string on the screen materializes the present state of the interpretation – as it is recorded by ABI. Two windows are presented to the histologist. The first (IMAGE BIOPSY 10) materializes a gray-level image together with a set of tools which help its reading. The second (STRUCTURE TYPE LIST) materializes the list of the names of characteristic pattern types, known to ABI. Note that the strings of characters on the screen are built as sets of pixels, i.e. as structures.

From the histologist's point of view, the interpretation is a two-step activity. First, s/he has to select the type of patterns s/he intends to look for. To this end, s/he points a name in the STRUCTURE TYPE LIST by a first mouse gesture. Then s/he has to identify the patterns of this type present in the image. For each pattern, s/he steers the mouse with a continuous gesture to draw the pattern closed boundary on IMAGE: BIOPSY 10. ABI, on its side, is a logbook, a set of visual sentences called *pages*. Each page description is a program P. Its *int* captures the pixels selected by each user gesture. P interprets the pixels according to its present state (i.e. the knowledge of the actions performed in the past and of the obtained results).

In the first interaction the message from the user to ABI is a single event, deriving from a single click on the mouse. P interprets this event to derive the type of the structures which will be selected next, the rules for their storage and representation and the line pattern for marking its instances on the biopsy image. In the second interaction, the message is a set of events identifying the pattern to be marked on the image. P interprets it deriving the coordinates of the pixels in the pattern contours and processing these data according to the current rules. The tools implementing *mat* materialize them as shown in Figure 11.1.

11.3.2 Image analysis, synthesis and visual reasoning revisited

The role of a *vs* and its components along a computation varies depending on the systems' current state, goals and available tools.

Image analysis: in a given context, a visual sentence *vs* is deduced from a given image *i* by a pattern recognition activity. The characteristic patterns present in the image are detected and described; the *vs* is completed by adding the definition of *int* and *mat*.

Image synthesis: in a given context, a visual sentence *vs* is deduced from a given description *d* by a synthesis activity and its pictorial part is materialized via both the synthesis and materialization rules holding for that context. The characteristic patterns are materialized depicting their meaning; firstly a bidimensional string is derived from d by applying synthesis rules and then *mat* materializes it on a suitable medium, e.g. a screen.

Visual reasoning: a *vs* supports visual reasoning and related communication; a set of pre-defined *vss* allows the user to describe and achieve the task goals, possibly by generating new *vss*.

11.4 VISUAL LANGUAGES

We are now in the position to formalize the intuitive notion of visual language. We are here interested not only in highlighting the mathematical structure of the *VL*, but also in specifying the different image and description domains which are necessary to design programs for visual computing. Last, we are interested in allowing users to specify *VL*s tailored to their interests; hence we shall examine the problem of the definition of *VL*s from the three different points of view. Through these three characterizations, visual languages can be regarded as a generalization of 1D languages (Salomaa, 1973). From the first point of view, we attempt to capture the formal structure of a visual language as a (possibly infinite) set of interpreted bidimensional strings.

Definition 12a   A *visual language VL* is a set of visual sentences:

$$VL \subseteq \{vs \,|\, vs = \langle i, d, \langle int, mat \rangle \rangle\}.$$

For example, interacting with ABI a biologist uses a visual language constituted by all the arrangements of windows, buttons and string of texts which can be interpreted as the page of a logbook in the biomedical domain. Figure 11.1 displays the image part of one of these visual sentences while Figure 11.3 shows the four components of a sub-sentence of Figure 11.1. Unfortunately, this definition neither can be considered as an operational one, nor does it clarify the complex relations among images and descriptions.

As a second point of view, we therefore adopt the *VL* designer one. The designers need to define programs that: a) analyse a given image to obtain an interpretation; b) synthesize an image from a given description; c) assess images exploiting the relations among a given image and a given description and d) use visual sentences in user-program dialogue. Hence, they are interested in a definition identifying the set of images and descriptions which are the objects and the targets of the image analyser and synthesizer, as well as the set of functions relating characteristic patterns in the images and substrings in the descriptions. It was pointed out that the 2D strings (images) of a visual language constitute a pictorial language *PL* and that the set of meaningful descriptions in the current operational context form a description language *DL*. A third component $\langle RE \rangle$ must be specified: a set of the pairs of functions $\langle int, mat \rangle$.

The following definition summarizes the latter considerations:

Definition 12b   A *visual language VL* is a triple $VL: \langle PL, DL, RE \rangle$. Each of the three sets can be defined following the technique better suited for the development style adopted by the system designer. This definition also allows a first classification of visual languages, based on the meanings associated with the images, i.e. on the description language *DL*. If a 2D string *i* in *PL* is regarded as the description of an activity to be performed – i.e. the characteristic patterns in *i* are interpreted as (sub-)programs – then, *VL* is a Visual Programming Language. If a 2D string *i* in *PL* is regarded as the representation of a scene – i.e. the characteristic patterns in *i* are interpreted as constants, then *VL* is a Visual Interpretation Language (Figure 11.5).

The third point of view we introduce is that of the users. For them it is useful to reason in terms of elementary visual sentences which can be composed to form complex ones.

For example, in Figure 11.1, a histologist reasons in terms of cells which are arranged to form a liver tissue. Hence, from the point of view of the users a visual language is defined by:

1)   A Visual Lexicon *VX*, a finite set of elementary visual sentences, i.e. a set of elementary images and associated computational meanings. For example, Figure 11.6 shows the lexicon of a visual programming language for the simulation of liver cell populations (Mussio *et al.*, 1991b). In this case the elementary images are the icons shown in Figure 11.6, the computational meaning of each icon being an object.

2)   A finite set *R* of visual composition rules, which state how to compose elements of the lexicon into complex visual sentences. Each rule is expressed by

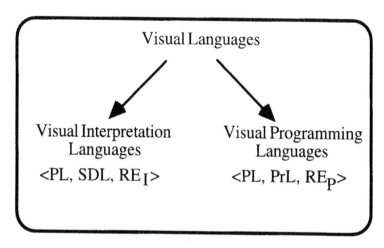

Figure 11.5   A classification of visual languages according to the description language used.

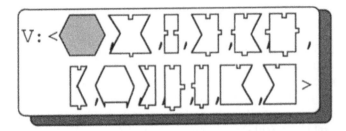

Figure 11.6 The set of elementary images in the cell visual lexicon.

a syntactic part and by a semantic one. The syntactic part specifies how to compose the pictorial sentences of the lexicon into the pictorial part of a new visual sentence; the semantic part how to derive the meaning of the new visual sentence from the meanings of the component images. Figure 11.7 shows the syntactic and semantic parts of a rule used to define the liver simulation

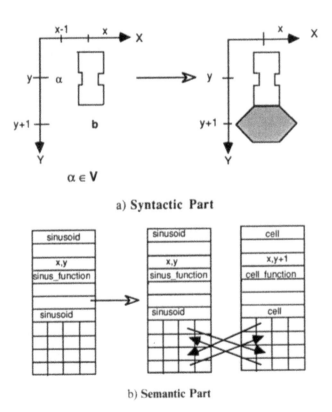

Figure 11.7 a) Syntactic and b) semantic parts of a visual rule. The rules implicitly define the relations between graphical elements in the syntactic part and elements in the descriptions, establishing the links and the values assumed by some object variables.

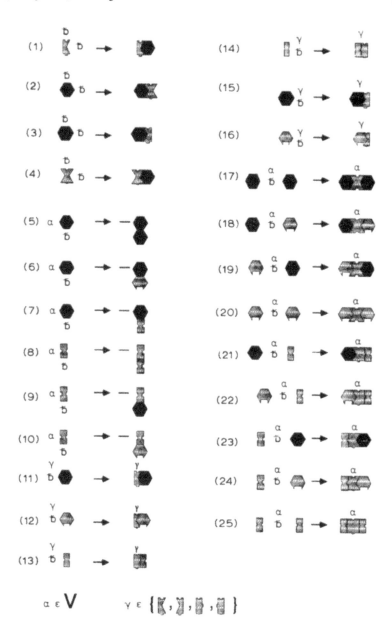

Figure 11.8 The syntactic rules of a visual language for the simulation of cell populations [reproduced from Mussio *et al.* (1991b). Reprinted with permission from Academic Press Ltd].

language. Figure 11.8 shows the syntactic parts of all the rules for that language.
3) The set of admissible visual sentences the users can reach by applying the
rules. This set is called the set of axioms (AX). In the case of the liver simulation
language, from the point of view of the users, the set of axioms is the set of
rectangular configurations obtained fitting together elementary images, i.e.
images which correspond to a correct puzzle as that shown in Figure 11.9.
Due to the definition of the semantic part of the rules, these configurations
correspond to executable programs.

The definition of *int* and *mat* results from the definition of VX and R. In this
way a) the shape and topology of the resulting structures appearing on the
screen are made explicit by the syntactic part of the rules (Figure 11.7a); b) the
evaluation of the semantic part of the rules establishes the geometrical properties of the structures as well as the links between the programs implementing *int*
and *mat* and the objects in the description (Figure 11.7b).
    Thirdly, a definition of visual language oriented towards the user is:

Definition 12c   A *visual language VL* is the set of *vss* generated (recognized) by
the system $\langle VX, R, AX \rangle$ where VX is a finite lexicon, R a set of visual
composition rules and AX a set of visual sentences on VX.
    Note that in the case of liver simulations, the set of axioms correspond to the
set of executable programs which can be represented as correct puzzles; from
the point of view of formal languages our definition allows the solution of the
membership problem through the construction of an interactive, incremental
compiler for the cell simulation language. However, the latter definition is quite
general: axioms can be defined to be the initial (final) state of a generative
(recognizing) procedure.

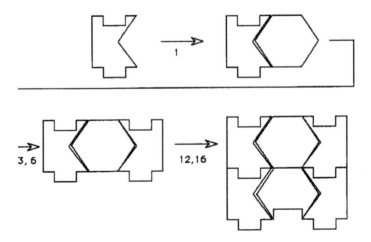

Figure 11.9   The derivation of a visual sentence.

11.5 A WALK THROUGH EXISTING SYSTEMS BASED ON VISUAL LANGUAGES

We explore the evolution of the visual language paradigm by reviewing five systems (VIVA, Khoros, LabVIEW, Smart Images, Image Interpretation Management Systems) which are devoted or can be used for image interpretation tasks. The five systems represent different phases in the evolution of the concept of visual language, even if their design and development occurred within the same time span. Note that the term 'visual language' is here used as defined in previous sections and not with the meanings that the different quoted authors associate with it.

The descriptions of these systems focus on different issues concerning user program interaction, visual programming and communication. Often, two different system presentations focus on rather different issues and even when they focus on a same issue, they reflect different design choices. Hence we try to standardize the different descriptions by assuming the following dimensions.

1) Parameters which influence the visual language design:
   a) the design goal;
   b) the design metaphor.
2) Parameters which define the visual language:
   c) visual lexicon;
   d) visual syntax and semantics.
3) Parameters which influence the user-program communication power:
   e) programming paradigm;
   f) programming style, granularity of the translation and execution mode;
   g) adaptability;
   h) liveness;
   i) reactivity.

The terms in (e) and (f) are used with the usual meaning in computer science. *Adaptability* is the capability of redefining or extending lexicon, syntax and semantics. *Liveness* (Tanimoto, 1990) is the level of live feedback the *vs* presents to its users. Tanimoto identifies four levels of liveness:

1) 'informative' level: a *vs* is a non-executable, passive visual document;
2) 'informative and significant' level: a *vs* is an executable visual document (e.g. the executable specification of a program);
3) 'edit-triggers updates' level: any edit operation performed by the user on the *vs* triggers a computation by the system;
4) 'stream-driven updates' level: the system continually updates the *vs* to show the generally time-varying results of processing streams of data according to the program as currently laid out.

Reactivity (Chang *et al.*, 1992) is the ability of the visual program to react to environmental changes or to the user.

VIVA   Visualization of Vision Algorithms (VIVA) (Tanimoto, 1990) provides a first frame for visual interactive image processing, underlining the importance of the feedback of the system toward the user. VIVA is an extended data-flow visual language for image processing. VIVA both provides a learning tool for students of image processing, and achieves a software foundation that can serve as a research platform for work on user-program interaction, parallel algorithm design, automatic or semi-automatic programming, attractive presentation medium for image processing and computer vision algorithms. Programs are defined following the metaphor of electronic circuits: components are connected by directional wires which determine the data flow (Figure 11.10).

The lexicon is constituted by two types of elements: nodes in the form of icons and one-directional wires in the form of arrows. Four types of nodes exist: sources (input programs), operators (image processing programs), monitors (output programs), control panels (I/O operators to be attached to programs to facilitate interactive checks). Each icon denotes a program. The syntax is not explicitly stated; wires and nodes are always active and carry their information simultaneously, ready to respond to triggers and to display the results of actions.

A program is a *vs*, built-in a 2D programming space by arranging the

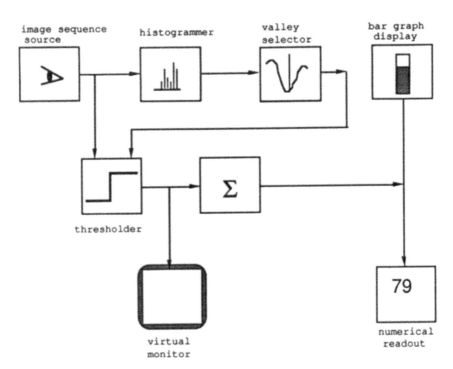

Figure 11.10   VIVA: block diagram of a visual program for image thresholding.

processing components and their interconnections. Procedural abstraction is provided by allowing the definition of icons denoting subprograms (Figure 11.10). Separate menu spaces are reserved for menu components and/or subprograms. A third type of space may be used by individual components to provide for special interaction with the user.

VIVA is designed to exhibit liveness at all four levels. The capability of redefining or extending lexicon, syntax and semantics is not mentioned. VIVA does not present reactive characteristics.

Khoros–Cantata   Khoros (Rasure and Williams, 1991) is designed stressing the integration of different tools in image processing together with the interface.

Khoros is a comprehensive system which relies on existing standards (X-Window and UNIX) to allow non-professional programmers to tailor it into high level interfaces to domain specific operators. A resulting Khoros application consists of domain-specific processing routines which are accessed by an automatically-generated user interface. The Khoros infrastructure consists of several layers of interacting subsystems whose organization is inspired to the metaphor of a chorus, 'any speech, song or other utterance made in concert by many people'. Khoros is based on Cantata, a general purpose visual multistyle data-flow programming language. The styles employed are flow-graphs, forms with various buttons and sliders, and textual programming.

A Cantata program is a data-flow graph, constructed from a user definable lexicon composed of: glyphs (operator or function), direct arcs (a path over which data tokens flow) and forms (interfaces between the visual data-flow graph and the textual nodes) (Figure 11.11). The descriptions of the lexicon elements are conventional programs. A textual label in the image part of a glyph identifies the associated description, arrow buttons identify the inputs and outputs, and icon buttons (e.g. Run, Reset buttons) are used to drive the execution mode (Figure 11.12).

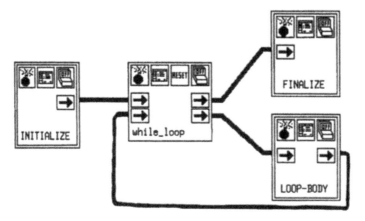

Figure 11.11   CANTATA: an example of a control structure.

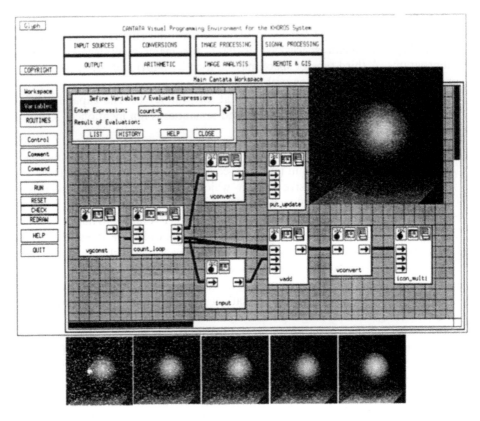

Figure 11.12 CANTATA: a noise cleaning visual program [reproduced from Rasure and Williams (1991), with kind permission from Academic Press Ltd].

A program is built by arranging the processing elements and their inter-connection. Procedural abstraction is allowed: a subgraph can be associated with a glyph and substituted by it in the main graph.

Cantata is an interpretative language: the user can stop the execution, modify parameters or connections and continue the execution or SAVE the entire workspace, including the data flow graph and intermediate results. It supports both data-driven and demand-driven execution approaches. It also supports single step execution, parallel execution and responsive modes. Khoros exhibits a) high user-driven adaptability; b) the fourth degree of liveness through animation of the glyphs; a programmable level of reactivity allowing automatic triggering of re-execution of the flow-graph. Originally based on a software environment created for research in image processing, Khoros is currently being used as a research and development tool to solve a variety of problems at numerous sites through the world. It is available at no cost via anonymous ftp at chama.unm.edu.

LabVIEW LabVIEW (National Instruments, 1992) is a commercial product, embodying some of the features of the first two systems, and including the capability of interacting with data acquisition and actuation tools. It is a graphical programming system for data acquisition and control, data analysis and data presentation. LabVIEW offers an object-oriented programming methodology in which the user graphically assembles software modules called *Virtual Instruments* (VIs). The exploited metaphor is the control panel of the VIs.

The lexicon has two types of elements: components of a Front Panel (user interface) and block diagram components to build the VI source code.

The elementary *vss* in a block diagram of a front panel are icons that can be also defined by the user exploiting an icon editor.

The elementary *vss* in a block diagram are: I/O components (routines to acquire or present data), computational components (routines to process data) and icon/connector components (to connect the input/output of VIs to front panels), lines to connect the different components (directing the flow of data). The data types (which will flow along the lines) are visually presented by the line thickness and/or color. Each type of component of a block diagram shows a different characteristic pattern. For example, the characteristic pattern of a logical operator is a rectangle with a label identifying the operator. The characteristic patterns associated with the control constructs are graphical structures surrounding the icons to be controlled (Figure 11.13).

A block diagram is built by placing components on the panel and connecting

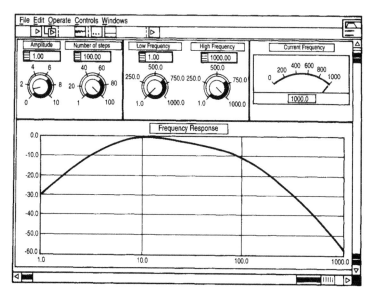

Figure 11.13 LabVIEW: a control panel [reproduced from National Instruments (1992), with kind permission from National Instruments].

the nodes of a block diagram by clicking on the output bottom of an icon and on the input button of another icon. As a response, LabVIEW connects the two points by a line. During the construction of a program, type checking is performed. If the input to an operator is not of the appropriate type, a warning is sent to the user.

Control structures are iterative loops, conditional loops and case statements for sequential, repetitive or branching operations.

The presence of modularity in LabVIEW allows procedural abstraction: a set of VIs can be associated with an icon and used in the block diagrams of other VIs. The single VIs can be tested alone.

LabVIEW is based on a data-flow paradigm, thus being able to simultaneously execute independent data paths and VIs. The underlying visual language is of an interpretative nature, but can also be compiled so as to obtain an optimized code.

An icon editor is available to the user to complete and refine the lexicon. Icons so defined can be used to build front panels or to associate to a subgraph. LabVIEW presents the second level of liveness, i.e. 'informative and significant' in that the block diagram specifies the computation and can be executed, but it does not present any reactive characteristics.

Smart image information system   Back to academic prototypes, within the biomedical domain the Smart Image Information Systems (SIIS) (Chang *et al.*, 1992) uses images not only as the objects to analyse but as a support to anticipate the user needs by accomplishing operations in advance without explicit requests. The SIIS exploits the data flow and rule-based paradigms to link an image to a knowledge structure.

The SIIS proposal is based on the metaphor of the 'smart house', where the house will automatically turn on lights, turn off the air conditioning system, switch the TV to monitor the entry door, and so on, according to environmental conditions and its design rules. In general, a smart house knows when to apply operations without an explicit request from the residents. Similarly a SIIS automatically responds to environmental changes by browsing, retrieving and pre-fetching or storing images. These reactions are guided by the knowledge structure S associated with an image I. The couple (I, S) image-knowledge structure is called 'smart image'. S determines the representation on the screen of data sets other than I in the form of graphs, sketches and different images.

This data structure can be profitably described as a *vs* composed of four types of visual sub-sentences: raw image, hot spot, protocol, and attributes; *int* and *mat* of these sentences are not made explicit in the definition of SIIS (Figure 11.14).

Raw images are *vs*s whose characteristic pattern is an image and whose descriptions are the programs for its management.

A *hot spot* is a *vs*, of which the characteristic pattern *ss* is a rectangular area

Protocol   **Image Database**   **Active Indexes**

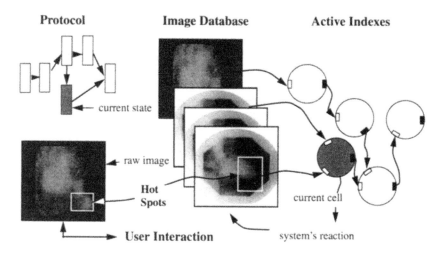

Figure 11.14   SMART IMAGE: overall structure [reproduced from Chang *et al.* (1992), with kind permission from Academic Press Ltd].

superimposed on a raw image. A hot spot description is a couple (p, a); p is a predicate that expresses a condition to be satisfied to trigger the action a, i.e. a sequence of activities. A hot spot is activated when an event associated with *ss* occurs, such as selection by the user or by a running program. Hot spot descriptions are organized as active indexes. An *active index* is a network of active cells. Each cell implements a decision rule, based on a variable number of inputs. Active indexes realize the system's responses to environmental changes.

A *protocol* codes and mimics the expected user's actions with/on the image. A protocol is derived by a visual lexicon of two types of elementary *vs*s: nodes and arcs. The characteristic pattern of a node is a rectangle having a string inside. Each node corresponds to a user's operation that causes environmental changes, and its description points to active index structures and/or hot spots. The rectangle background can be in one of two colors, signalling if the hot spot is active or not. Arcs are represented by arrows and determine the data flow. Hence the image part of a protocol results into a flowchart of named rectangles linked by arrows. The protocol description is specified by a tuple (protocol name, decision-graph, protocol-state). The protocol name is the unique identifier of a specific protocol, the decision graph describes the strategy of image usage in a certain application and the protocol state is the current state in this strategy.

Attributes are domain-related information and traditional image information (e.g. patient name, image size, ...) visualized as formatted text. The user can exploit editors to define and update a smart image with respect to each of its components. Moreover, a smart image incrementally increases its intelligence by updating its knowledge structure to include more information on the image usage. To allow this, the lexicon of the SIIS visual language is enriched by hot spot buttons (to associate a hot spot description to a hot spot image), hot spot definition buttons (to define new hot spot types) and command buttons (associated with operations such as quit and voice recording).

The syntax of the visual language is not mentioned.

SIIS presents the 'informative and significant' second level of liveness: hot spots describe the computation to be performed. Moreover, it is adaptable in that the user can define new hotspots.

In conclusion, SIIS is: a) proactive, in that environmental changes caused by environment actions may automatically trigger further actions; b) reactive, in that environmental changes caused by users' interaction with images can trigger some actions; c) reflective, in that actions activated by conditions on hot spots can change features on these hot spots.

Image Interpretation Management System The Image Interpretation Management System (IIMS) (Bottoni *et al.*, 1993) is an instance of the Cooperative Visual Environment (CVE) (Bianchi *et al.*, 1993) focused on image interpretation.

A CVE operates with its users in performing some situated activity exploiting the metaphor of their real working environment. It can also be instrumented to a specific task by the virtual tools users foresee as useful, as happens in real working environments. A user interface development environment embedded in the CVE permits the users to evolve its organization, the interaction mode and the tools as soon as they gain deeper insight on their activity. A CVE is proactive, in that environmental changes caused by program actions may automatically trigger more actions: for example the loading of an image can cause the computation of its histogram and the derivation of a segmented image. It is also reactive in that environmental changes caused by users' interaction with images can trigger some actions: for example zooming an image can cause to zoom another image logically related to the first one. It also exhibits liveness at all four levels: a CVE is continually active and its users can, at a certain state of a program execution, modify it and resume its execution from that state.

The IIMS inherits this structure and lets its users directly customize the image interpretation virtual system to their real working environment. For example, customization of the system is obtained by extending traditional notations used by physicians for biopsy interpretation and documentation.

Two different visual languages have been defined, the Documentation Language (*DL*) and the Image Interpretation Language (*IIL*). Their visual sentences are shuffled to form the actual IIMS ones, as shown for example in Figure 11.1.

Both *DL* and *IIL* are formally defined following Definition 12c: this definition is used in the case of the documentation language to develop an incremental compiler which manages the interaction. In the case of the interpretation language, it is exploited to incrementally interpret the images of a given class.

DL sentences are built following the metaphor of the logbook that physicians use in real working environments to log their clinical activity. The result is a computerized logbook (c-logbook) organized, as a traditional one, in pages (c-pages); these pages are in turn displayed on the screen one at a time inside a fixed frame: the c-cover. Note that c-cover, c-page, c-logbook are *vs*s whose image parts are close to the characteristic patterns of medical documentation (Mussio *et al.*, 1992).

The *IIL* lexicon elements (nuclei and other anatomical structures) are the characteristic patterns by which a physician describes a biopsy interpretation and the *IIL* composition rules derive from those followed empirically in human interpretation. However, the *vs*s in the two visual languages play different roles in the interaction: c-logbooks are implemented to reason on images. The c-logbook definition is mainly used to generate surface images and to control the user-program interaction. The *IIL* definition is used to (semi)-automatically interpret an image, create a *vs* from the original image and the obtained description and drive the *vs* interaction with the users (Bianchi *et al.*, 1991; Bottoni *et al.*, 1993; Bottoni *et al.*, 1995).

An academic prototype of the IIMS has been implemented and customized for several medical applications (Bianchi *et al.*, 1991; Bottoni *et al.*, 1993).

11.6 CONCLUSIONS

A unified language for the definition of image processing, interpretation, reasoning and synthesis, based on the novel concept of *visual sentence* is offered as a proposal for discussion. The significance of this unified language is twofold. First, it helps to avoid ambiguity in discussing on activities involving images as often happens in the literature. Second, it allows a uniform formal specification of all the activities occurring when interacting through the screen and hence it may become a tool for the comprehensive description of all visual computing activities.

A visual sentence relates an image and its characteristic patterns to their descriptions by making explicit the relations which map characteristic patterns into procedures and vice versa. The set of characteristic patterns, as well as the set of descriptions, are defined as formal languages.

This formal approach allows the capture of the dynamical aspects of reasoning with/on images. An image describes a program: the evolution of the computation is reflected along the image transformation. At a given instant, the image can completely and unambiguously describe the state of the computation. It also allows the exploitation of ambiguities, arising in image interpretation, to better grasp all the possible image meanings – which seems to be a common procedure within scientific activities. In the latter cases, a single image may be concurrently described from different sets of characteristic patterns. The different descriptions can thereafter be managed to reason – possibly automatically – on the image significance.

These advantages for reasoning on and with images have been illustrated examining different examples.

REFERENCES

Brancheau, J.C. and Brown, C.V. (1993) The management of end-user computing: status and direction. *ACM Comp. Surveys* **25**, 437–482.

Bianchi, N., Bottoni, P., Cigada, M., De Giuli, A., Mussio, P. and Sessa, F. (1991) Interpretation strategies in a cardiologist-controllable automatic assistant for ECG description. *Proc. IEEE Comp. in Card.*, 673–676.

Bianchi, N., Bottoni, P., Mussio, P. and Protti, M. (1993) Cooperative visual environments for the design of effective visual systems. *Journal of Visual Languages and Computing* **4**, 357–382.

Bottoni, P., Cugini, U., Mussio, P., Papetti, C. and Protti, M. (1995) A system for form feature based interpretation of technical drawings. *Machine Vision and Applications*, **8**, 326–335.

Bottoni, P., Mussio, P. and Protti, M. (1993) Metareasoning in the determination of image interpretation strategies. *Pattern Recognition Letters* **15**, 177–190.

Cugini, U., Falcidieno, B., Mussio, P. and Protti, M. (1988) Towards automatic indexing of product models in CIM environments. *Proc. 6th Int. Workshop on Languages for Automation*, 106–114.

Chang, S.-K. (1986) Introduction: visual languages and iconic languages. In: *Visual languages.* (eds S-K. Chang, T. Ichikawa and T. Ligomenides) 1–7. Plenum Press, New York.

Chang, S.K. (1995) Toward a theory of active index. *J. of Visual Languages and Computing*, **6**, 101–118.

Chang, S-K., Hou, T.Y. and Hsu, A. (1992) Smart image design for large data-bases. *J. of Visual Languages and Computing* **3**, 323–342.

Chang, S-K., Tortora, G., Yu, B. and Guercio, A. (1987) Icon purity – towards a formal decision of icon. *Int. Jour. of Pattern Recognition and Artificial Intelligence* **1**, 377–392.

Fu, K.S. and Tsai, W.H. (1980) Attributed grammars: a tool for combining syntactic and statistical approaches to pattern recognition. *IEEE Trans. on Syst., Man and Cybernetics* **SMC-10**, 873–885.

Golin, E.J. (1991) Parsing visual languages with picture layout grammars. *J. of Visual Languages and Computing*, **2**, 371–394.

Haralick, R.M. and Shapiro, L.G. (1991) Glossary of computer vision terms. *Pattern Recognition* **24**, 69–93.

Knuth, D. (1968) Semantics of context-free languages. *J. Math Syst. Theory* **2**, 127–145.

Kobsa, A. and Walster, W. (eds) (1989) *User models in dialog systems.* Springer-Verlag, Berlin, Germany.

Mussio, P., Finadri, M., Gentini, P. and Colombo, F. (1992) A bootstrap approach to visual user interface design and development. *The Visual Computer* **8**, 75–93.

Mussio, P., Pietrogrande, M., Bottoni, P., Dell'Oca, M., Arosio, E., Sartirana, E., Finanzon, M.R. and Dioguardi, N. (1991a) Automatic cell count in digital images of liver tissue sections. *Proc. 4th IEEE Symp. on Computer-Based Medical Systems*, 153–160.

Mussio, P., Pietrogrande, M. and Protti, M. (1991b) Simulation of hepatological models:

a study in visual interactive exploration of scientific problems. *J. of Visual Languages and Computing*, **2**, 75–95.

Narashiman, R. (1969) On the description, generation and recognition of classes of pictures. In: *Automatic interpretation and classification of images* (ed. A. Grasselli), 1–42. Academic Press, London, UK.

National Instruments Corporate Headquarters (1992) *LabVIEW for Windows, demonstration guide.*

Reiter, R. and Mackworth, A. (1989/90) A logical framework for depiction and image interpretation. *Artificial Intelligence* **41**, 125–156.

Rosenfeld, A. and Siromoney, R. (1993) Picture languages – a survey. *Languages of Design* **1**, 229–245.

Rasure, J.R. and Williams, C.S. (1991) An integrated data flow visual language and software development environment. *J. of Visual Languages and Computing* **2**, 217–246.

Salomaa, A. (1973) *Formal languages.* Academic Press, New York.

Shaw, A.C. (1969) The formal picture description scheme as a basis for picture processing systems. *Information and Control* **14**, 9–52.

Siegel, J. (1956) *Non-parametric statistics in social sciences.* McGraw-Hill, New York.

Tanimoto, S.L. (1990) VIVA: a visual language for image processing. *J. of Visual Languages and Computing* **1**, 127–140.

Tondl, L. (1981) *Problems of semantics.* Reidel, Dordrecht, The Netherlands.

Wang, D. and Lee, J.R. (1993) Visual reasoning: its formal semantics and applications. *J. of Visual Languages and Computing* **4**, 327–356.

Wittenburg, K., Weitzman, L. and Talley, J. (1990) Unification-based grammars and tabular parsing for graphical languages. *J. of Visual Languages and Computing* **2**, 347–370.

12

Visual Databases

12.1   INTRODUCTION

Databases of images and image sequences are becoming a subject of increasing relevance in multimedia applications. Archiving and retrieving these data requires different approaches from those used with traditional textual data. Specifically, the marked characterization of the information units managed in such systems, with the lack of a native structured organization and the inherent visuality of the information contents associated with pictorial data makes the use of conventional indexing and retrieval based on textual keywords unsuitable. Icons as picture indexes have been suggested to support image retrieval by contents from databases (Tanimoto, 1976), iconic indexes standing for objects, object features or relationships between objects in the pictures.

The use of iconic indexes naturally fits with visual querying by example to perform retrieval. In this approach, image contents are specified by reproducing on the screen of a computer system the approximated pictorial representation of these contents. Using pictures to query pictures, retrieval is thus reduced to a matching between the pictorial representation given by the user and the iconic indexes in the database. Visual queries by example make simpler the expression of image queries by contents. The cognitive effort which is required to the user in the expression of image contents through text is now reduced by resorting on the natural human capabilities in picture analysis and interpretation.

ARTIFICIAL VISION
ISBN 0-12-444816-X
Copyright © 1997 Academic Press Ltd
All rights of reproduction in any form reserved

In the following, we will review several of the main approaches in pictorial data iconic indexing and visual retrieval. Single image indexing and retrieval is discussed in Section 12.2, with reference to different approaches to indexing and retrieval based on different facets of the informative contents of pictorial data, such as spatial relationships, color, textures and object shapes. The subject of sequence retrieval is discussed in Section 12.3 considering symbolic indexing and iconic retrieval.

12.2 SINGLE IMAGES

Pictorial indexing and querying of single images may take into account relative object spatial relationships, color and textural attributes, or shapes of the objects represented in the images.

12.2.1 Iconic indexing and querying based on spatial relationships

Spatial relationships between imaged objects play the most immediate and relevant role for the indexing of image data. The symbolic projection approach, initially proposed by Chang and Liu (1984) and Chang *et al.* (1987), and subsequently developed by many authors (Lee *et al.*, 1989; Chang and Jungert, 1991; Costagliola *et al.*, 1992; Jungert, 1992; Lee and Hsu, 1992; Del Bimbo *et al.*, 1993a, 1994a), is a simple and effective way to provide iconic indexes based on spatial relationships and is also ideally suited to support visual interaction for the expression of queries by contents. In this approach, first, after preprocessing, the objects in the original image are recognized and enclosed in their minimum bounding rectangles. Then for each object, the orthogonal relation of objects with respect to the other objects are created. Objects are abstracted with their centroids. Orthogonal projections of the object centroids on the two image axes x and y are considered. Three spatial relational operators $\{<, =, :\}$ are employed to describe relations between object projections. The symbol '$<$' denotes the left–right or below–above spatial relationship, the symbol '$=$' denotes the at the same spatial relation as, the symbol '$:$' stands for in the same set as relation.

Image contents may be described through a *2D string* (Chang and Liu, 1984). A 2D string (u, v) over a vocabulary of pictorial objects o_i, is formally defined to be: $(x_1 \; op \; x_2 \; op \; x_3 \; op \; x_4 \; op \ldots, y_1 \; op \; y_2 \; op \; y_3 \; op \; y_4 \; op \ldots)$, where op is one of the relational operators expounded above and x_i and y_i are the projections of the objects o_i along the x- and y-axes, respectively.

A pictorial query can be expressed visually, by arranging icons over a screen to reproduce spatial relationships between objects and in turn reduced to a 2D string. In this way, the problem of pictorial information retrieval is reduced to the matching of two symbolic strings. Examples of visual retrieval according to the symbolic projection approach using 2D string encodings are reported in (Chang *et al.*, 1988).

Among the developments of the original 2D-string approach, *2D G-strings* with the *cutting mechanism* (Lee *et al.*, 1989) overcome the limitations of 2D strings due to the inability to give a complete description of spatial relationships that occur in complex images. A 2D G-string is a five-tuple $(V, C, E, e, \langle - \rangle)$ where V is the vocabulary; C the cutting mechanism which consists of cutting lines at the extremal points of the objects; $E = \{<, =, !\}$ is the set of extended spatial operators; e is a special symbol which represents an area of any size and any shape, called the *empty space object*, and '$\langle - \rangle$' is a pair of operators which is used to describe a local structure. The cutting mechanism defines how the objects in an image are segmented. The special empty space symbol and the operator pair '$\langle - \rangle$' provide the means to use generalized 2D strings to substitute for other representations.

Although 2D G-strings can represent the spatial relationships among objects in pictures, the number of segmented subparts of an object is dependent on the number of bounding lines of other objects which are completely or partially overlapped. Lee and Hsu (1992) define *2D C-strings*, which preserve all spatial relations between objects with efficient segmentation. The knowledge structure of 2D C-string is a five-tuple $(S, C, Rg, Rl, ())$, where: S is the set of symbols in symbolic pictures of interest; C is the cutting mechanism which consists of cutting lines at the points with partial overlapping from x- and y-projection respectively. Cuttings are performed along the x- and y-axes independently, in a way that the 2D-C-string representation of a picture is unique and minimal. The set $Rg = \{<, !\}$ is the set of global relational operators; considering two objects A and B operated with the above operators, they have the meaning of A *disjoins* B and A *is edge to edge to* B, respectively; $Rl = \{=, [,], \%\}$ is the set of local relational operators considering two objects A and B operated with the above operators, they have the meaning of A *is the same as* B, A *contains* B *and they have the same begin-bound*, A *contains* B *and they have the same end-bound*, A *contains* B *and they have not the same bound*, respectively; () is a pair of separators which is used to describe a set of symbols as one body.

An extension of 2D strings to deal with *three-dimensional imaged scenes* was proposed by Del Bimbo *et al.* (1993a). The approach relies on the consideration that two-dimensional iconic queries and 2D-string-based representations are effective for the retrieval of images representing 2D objects or very thin 3D objects, but they might not allow an exact definition of spatial relationships for images representing scenes with 3D objects. In fact in this case, an incorrect representation of the spatial relationships between objects may result, due to two distinct causes. First, 2D icons cannot reproduce scene depth. 2D icon overlapping can be used only to a limited extent since it impacts on the intelligibility of the query. Second, as demonstrated by research in experimental and cognitive psychology, mental processes of human beings simulate physical world processes. Computer-generated line drawings representing 3D objects are regarded by human beings as 3D structures and not as image features, and they imagine spatial transformations directly in 3D space. Therefore an unambiguous correspondence is established between the iconic query and image contents if the spatial relationships referred to are those between the objects in the scene represented in the image, rather than those between the objects in the image. The dimensionality of data structures associated with icons must follow the dimensionality of the objects in the scene represented in the image. A 3D structure should be employed for each icon to describe a 3D scene. Representations of images are derived considering 3D symbolic projections of objects in the 3D imaged scene. Thirteen distinct operators, corresponding to the interval logic operators distinguish all the possible relationships between the intervals corresponding to the object projections on each axis.

Retrieval systems employing the ternary representation of symbolic projections have been expounded in Del Bimbo *et al.* (1993a) and Del Bimbo *et al.* (1994a). In the latter approaches, the user reproduces a three-dimensional scene by placing 3D-icons in a virtual space and sets the position of the camera in order to reproduce both the scene and the vantage point from which the camera was taken. A spatial parser translates the visual specification into the representation language and retrieval is again reduced to a matching between symbolic strings.

12.2.2   Iconic indexing and querying based on colors and textures

Indexing and query by contents based on picture colors or object textures has been proposed, among the others, by Hirata and Kato (1992), Binaghi *et al.* (1992) and Niblack *et al.* (1993). Generally speaking, image color distribution and textures are assumed to be the characterizing features of image contents and images are requested that contain objects whose color/texture is similar to the color/texture selected from some user menu, or that have global similarity as to color or textures.

In the QBIC system (Niblack *et al.*, 1993) color space is represented by using average (R, G, B), (Y, i, q), (L, a, b) and the *Mathematical Transform to Munsell coordinates* (MTM) of each object. A color histogram is also evaluated. Color RGB space is quantized in 4096 cells. The MTM coordinates of each cell are computed and clustering is performed to obtain the best 256 colors. Color histogram is obtained as the normalized count of the number of pixels that fall in each of these colors. For average color, the distance between a query object and the database object is weighted Euclidean distance. Improved results are claimed when color matching is not made on average color but on distribution of colors occurring in an object or image. Queries representing images with distinct percentages of different colors are also allowed.

A similar approach is carried out by Hirata and Kato (1992). Iconic indexes are built following a complex procedure. Images are divided into several regions using edges (see the next section), color values of the pixels, texture values. The edge-detection step uses an adaptive differential filter in the RGB space based on the Weber–Fechner law of human vision. In the color measurement the image is divided into regions considering the distance in the hue, lightness and saturation space. Color and texture values are added to each region. Retrieval is obtained by measuring the similarity (through *correlation functions*) of shapes (see the next section), color and position between regions in the user sketch and the iconic indexes in the database.

12.2.3 Iconic indexing and querying based on object shapes

Several mathematical frameworks can be defined to measure shape similarity. The most straightforward way is to extract a number of features describing the shape and then make a number of measurements to determine the distance in the feature space between the *model* (the shape we draw) and the *candidate* (the part of the image that we are testing for the presence of the shape). Nevertheless, unlike other indexes like color or texture, shape does not have a mathematical definition that exactly matches what the user feels as a shape. It is not easy to understand what a human perceives as shape and, more importantly what should be considered a valid shape similarity measure. In particular, it often happens that shapes that a human feels as very similar are regarded by reasonable, mathematically-defined distance operators as completely different. Feature based approaches tend to be brittle in the presence of shape distortion and generally speaking a feature-based comparison between a model and an image just doesn't work. Typical images span very high-dimensional subspaces in feature spaces, thus requiring the extraction of a great number of features to be reliably characterized, and, for most of these features, there is no warranty that our notion of *closeness* is mapped in the topological closeness in the feature space.

Retrieval by contents based on object shapes and sketches has been proposed by Hirata and Kato (1992) and in the QBIC system by the IBM research group in Almaden (Niblack *et al.*, 1993). In the latter work, similarity retrieval is carried out by considering the *local correlation* between a linear sketch drawn by the user on a tablet and an abstract image in the database. *Linear sketches* are binarized sketches which passed a thinning and shrinking procedure to be reduced to a fixed size sketch (64×64). *Abstract images* are images that are converted to a single band luminance, passed a Canny edge extraction procedure, a thinning and shrinking procedure to be reduced to the standard 64×64 size. In the correlation process a local block is shifted in 2D directions to find the best match position between the sketch and the image.

In the QBIC system (Niblack *et al.*, 1993), heuristic shape features are considered such as *area* (i.e. the number of pixels set in the binary image), the *circularity*, computed as the square perimeter over area, *major axis orientation*, computed as the direction of the largest eigenvector of the second order covariance matrix, and *eccentricity*, measured as the ratio between the smallest and the largest eigenvalues. *Algebraic moment invariants* are also considered that are computed from the first m central moments. Shape matching between the sketch and stored images is accomplished by considering a weighted Euclidean distance between the features in the two images.

Limitations of both the above approaches stand in the requirement that shapes drawn by the user must be close to the imaged shapes of the objects in the requested images, due to the rough parameterization used. Moreover, due to the distance function used, given a certain shape many images other than that containing the requested object are answered, thus forcing the user to perform a further analysis over a still large set of images. Since retrieval is only based on shapes, without any care of spatial relationships, in the case of images with multiple objects, the lack of pruning over the image database may lead to unmanageable sets of answered images.

Del Bimbo *et al.* (1994b) present a system for image retrieval by contents, following a different approach, based on shape matching with *elastic deformations*. In this system, the user sketches a drawing on the computer screen and the drawing is deformed to adapt to shapes of objects in the images. Differently from previous attempts in shape-based image retrieval, similar shapes are searched with no form of feature extraction.

In this system, images are collected and, while inserted in the database, a number of *interesting rectangular areas* are selected. Typically, these areas correspond to objects in the image. All the following searches are limited to the interesting areas. A query is again composed by drawing a sketch of one, or more, shapes on a graphic screen. A *template*, in the form of a 4th-order spline, is instantiated for each of the shapes. Image retrieval is based upon two considerations: a candidate image is retrieved and presented for browsing if it has two or more areas of interest in the same spatial relationship as the shapes drawn on the tablet and the shapes contained in the areas of interest match the shapes on the tablet with a certain degree.

To make a robust match even in the presence of deformations, we must allow the template to deform. The template must deform taking into account two opposite requirements: first, it must *follow as closely as possible* the edges of the image (R_1); second, it must take into account the deformation of the template (R_2): specifically a measure must be taken into account of the energy spent to *locally stretch and bend* the template. The elastic energy depends only on the first and second derivatives of the deformation. This allows not to penalize discontinuities and sharp angles that are already present in the template, but to penalize only the degree by which we depart from those discontinuities or angles. Also, since the energy depends only on the derivatives of deformation, pure translation of the template, for which deformation is constant, does not result in additive cost. This makes the scheme inherently translation invariant. The goal is to maximize R_1 while minimizing R_2. This can be achieved by minimizing a compound functional where the term R_1 appears with plus sign and the term R_2 with minus sign. A solution can be obtained as a solution of the Euler–Lagrange equations associated to this variational problem. Under suitable approximations the solution is a spline function with i knots. The match between the deformed sketch and the imaged objects, as well as the deformation energy of the sketch are taken into account to measure the similarity between the model and the image. Energy and match values at the end of the deformation process, together with a measure of the shape complexity (the number of zeroes of the curvature function) are used as an input to a neural classifier which derives a similarity ranking. Only images with the highest similarity rankings are displayed in response to the query. An example of retrieval by elastic matching is shown in Figure 12.1, where the rough sketch of a bottle is compared with images from the Morandi's catalogue: the deformed template is shown superimposed with the six *most similar* images in the database. For queries including multiple objects, spatial relationships between the different objects are taken into account.

The elastic matching algorithm is computationally very demanding, and sets a serious burden on the capacity of the system it is implemented on. To make this system as effective as possible, it is important to make a preliminary filtering

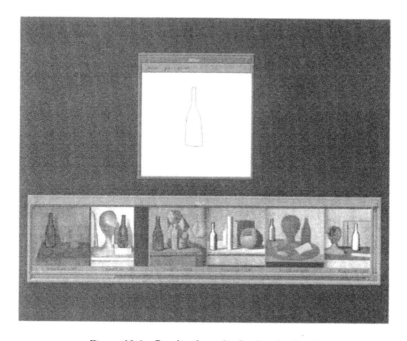

Figure 12.1   Retrieval results for bottle sketch.

of the database, eliminating as many images as possible before applying the elastic matching algorithm.

12.3 IMAGE SEQUENCES

As opposite to the number of experiences on single images, only a few techniques have been proposed for indexing and querying of image sequences. Contents of frame sequences can be expressed by referring either to the *temporal structure of the video*, or to the *occurrences in the individual frames*. The first approach especially makes evident the editing process which has been applied to create the syntactic structure of the video; the second, points out the *spatial occurrences of objects in the individual frames*, or the *temporal changes of spatial relationships*, or, at a higher level, the *spatio-temporal interactions among imaged objects*. Indexing through edit types is especially important for storage and retrieval of films and is based on special segmentation techniques that allow to detect boundaries of the syntactic units of the film. It will not be discussed here. In the following we will shortly review the indexing of sequence contents through high-level descriptions (Section 12.3.1), and through representation of spatio-temporal relationships between objects in the frame sequence (Section 12.3.2).

12.3.1 High-level indexing and retrieval

High-level interactions between entities are represented through *semantic networks* by Walter *et al.* (1992). Their study concentrates on a formalism for the symbolic representation of physical entities represented in the images, abstract concepts, their dynamic changes and a matching formalism for the representation of deductive rules. *Episodes* are extracted from a sequence and represented as ensembles of objects and processes. Objects are all material entities and properties or relations between material objects, while processes represent changes of properties and relations between objects over time. Abstract categories of objects are called *generic descriptions* and particular examples of object categories are referred to as *individuals*. Individuals may belong to several *genera* not necessarily at the same time. Genera may be connected by various relations through generalization or attribute links. The representation formalism is referred to as EPEX-F and is a variation of KL-ONE. EPEX-F includes a temporal logic to allow temporal information to be maintained with the objects as lifetime attributes. Retrieval is made textually and supported by a rule system. This approach supports description of episodes at a high semantic level; nevertheless its effectiveness is limited to a small number of objects in the episodes. Queries are easier to be expressed textually than visually due to the abstraction level of relationships between entities.

Davis (1993) indexes video sequences through *high-level episodes* reflecting the occurrence of typical situations, such as a person running, or drinking from a cup. Indexing is handmade and is obtained by associating a *situation icon* to each set of frames in which the situation occurs. To retrieve subsequences, icons can be combined visually to define even complex visual queries.

12.3.2 Iconic indexing and querying based on spatial relationships

Arndt and Chang (1989) propose the extension of 2D strings to the indexing of image sequences. Sequences are divided into a number of series where each series represents a continuous sequence of images. Sequence descriptions are obtained by deriving the 2D string for each set of frames where the description keeps constant. Changes that may occur between frames (and in the 2D strings altogether) are marked with the number of frame where changes occur. *Sets operators* such as *addition*, *deletion* of an object, *merging* and *splitting* of sets, *deletion* and *addition* of sets are used to identify the type of change occurred.

However, such a representation, while effective for indexing image sequences at a lower level with respect to semantic network approaches, does not provide the flexibility needed in representation of image sequences for their retrieval. Two types of problems must be addressed. First, the specificity of image sequence contents, which imposes constraints on the type of representation used. Second, the fact that querying image sequences must be supported at different levels of detail according to the actual knowledge of the user about sequence contents. The first issue requires representations that avoid ambiguities in the description of image contents; the second requires that a language is available supporting different levels of detail and refinement. As to the first issue, in the representation of contents of image sequences, we must take into account that image sequences usually represent 3D dynamic real-world scenes with three-dimensional motions of multiple objects. According to what was discussed for the case of single images, descriptions referring to the original imaged scene (*3D scene-based descriptions*) are generally needed to avoid representation ambiguities. Image-based descriptions can be considered only in the special cases where all the objects lay on a common plane and the camera is in a normal position with respect to it. In 3D scene-based descriptions, two different descriptions are possible, depending on the reference systems on which symbolic projections are taken. On the one hand, object projections can be evaluated with reference to the Cartesian coordinate system of a privileged point of view, corresponding to the vantage point of the viewing camera (*observer-centered* description). In this case, images of the same scene, taken from different viewpoints, are associated with distinct 3D scene descriptions. On the other hand, object projections can be evaluated with reference to the coordinate systems associated with individual objects (*object-centered*

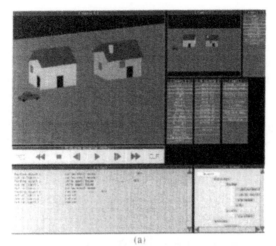

(a)

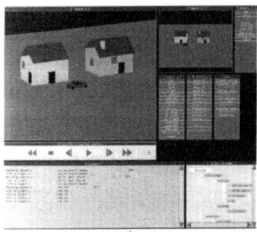

(b)

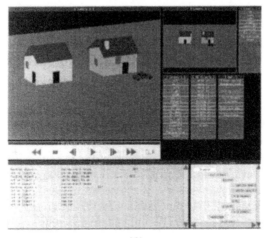

(c)

description). The overall description of the scene is obtained as the composition of multiple object-centered descriptions, each capturing how one object sees the rest of the scene. Since in the object-centered approach descriptions are independent of the observer point of view, images of a scene, taken from distinct viewpoints, are all associated with the same 3D description. Considering the evolution over time of scene-based descriptions, observer-centered representations lead to a sequence description which is dependent on the observer point of view. Object motion is correctly represented only if the camera is fixed. In the presence of a moving camera, objects may be associated with apparent motions. For instance, camera zooming results in expansive or contractive motions. As a consequence, the description associated with the sequence may include changes in spatial relationships that are not due to the actual motion of the objects. Whereas, object-centered scene descriptions always result in the representation of the actual motion of the objects, both with a fixed and with a moving camera. In this case, the sequence description does not depend on the observer point of view and only represents actual changes in the spatial relationships between objects as occurring in the original imaged scene.

These considerations have been expounded by Del Bimbo *et al.* (1993b), where a language is also proposed for the symbolic representation of spatio-temporal relationships between objects within image sequences is presented. This language, referred to as *Spatio-Temporal Logic* (STL), comprises a framework for the qualitative representation of the contents of image sequences, which allows for treatment and operation of content structures at a higher level than pixels or image features. STL extends basic concepts of *Temporal Logic* (Allen, 1983) and symbolic projection to provide indexing of sequence contents within a unified and cohesive framework. By inheriting from Temporal Logic the native orientation towards qualitative descriptions, this language permits (without imposing) representations with intentional ambiguity and detail refinement, which are especially needed in the expression of queries. Besides, by exploiting the symbolic projection approach, STL supports the description of sequence contents at a lower level of granularity with respect to semantic networks and annotation-based approaches, thus allowing to focus on the actual dynamics represented in the sequence. Temporal properties are expressed as Temporal Logic assertions that capture the evolution over time of spatial relationships occurring in single scenes. These relationships are expressed through an original language, referred to as *Spatial Logic*, which transposes the concepts of Temporal Logic itself to express relationships between the projections of the objects within individual scenes.

The spatial assertions that capture the properties of the contents of the individual scenes are composed through the Boolean connectives *and, or, not* and their derived shorthands, and the *temporal-until* operator *unt_t*. Boolean connectives have their usual meaning and permit the combination of

◄

Figure 12.2 Visual query by example. Specification of a simple dynamic scene: (a) The initial scene; (b), (c) specification of motion by the dragging of the car icon.

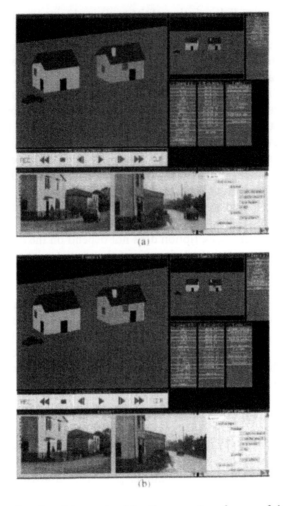

Figure 12.3 Results of the retrieval for the scene (two frames of the sequence).

multiple assertions which refer to any individual scene of the sequence. The temporal-until operator permits to define temporal ordering relationships between the scenes in which different assertions hold. Specifically, temporal until is a binary operator which permits the composition of a pair of assertions q_1 and q_2 to express that q_1 holds along the sequence at least until reaching a scene in which q_2 holds. Shorthands of the *unt_t* operator are the *temporal-eventually* operator (*eve_t*) and the *temporal-always* operator (*alw_t*). Temporal eventually is a unary operator which permits the composition of an assertion q_1 to express that somewhere in the future q_1 may hold (nothing is said about what holds until q_1 will hold). The temporal-always operator expresses that an assertion q_1 will hold from now in the future. Spatial Logic exploits the symbolic projection approach for the description of spatial relationships between objects. It transposes the concepts of Temporal Logic so as to capture the geometric ordering relationships between the projections of the objects within a multi-dimensional scene.

If Φ is a spatial formula defining some positional properties, the spatial assertion of the form $(S_E, q, e_n) := \Phi$ (read $\Leftarrow$ as models) where q is an object possibly extending in more than one region, expresses that the relational property Φ holds in any region containing the object q, considering projections along the e_n reference axis.

For instance, the assertion $(S_E, q, x) := p$ (which is referred to as *spatial alignment* with respect to the x-axis) expresses that the projection of q on the x-axis is entirely contained in the projection of p.

The assertion $(S_E, q, x) := (eve\_s + p)$ expresses that every point of the projection of q on the x-axis of the underlying reference system E has at least one point of the projection of p to its right side $(s+)$.

The assertion, $(S_E, q, x) := (q \ unt\_s + p)$ expresses that the projection of q on the x-axis extends at least until the beginning of the projection of p.

Composition of the above spatial operators through the Boolean connectives allows the expression of spatial relationships at different levels of detail, which is especially useful in the formulation of queries.

A retrieval system for which exploits such a language for iconic retrieval by example of image sequences is also presented in that paper. In this system sequences are stored with their STL symbolic descriptions associated. Queries are expressed by the user through a visual iconic interface which allows for the creation of sample dynamic scenes reproducing the contents of sequences to be retrieved. Sample scenes are automatically translated into spatio-temporal logic assertions, and retrieval is carried out by checking them against the descriptions of the sequences in the database. An example of retrieval of image sequences based on STL descriptions is shown in Figure 12.2. The pictures in the first column present different steps of the visual query specification (selection of the objects (a)) and motion definition (b), (c): sequences where a car is passing in front of two houses are searched. The pictures in Figure 12.3 show two frames of the retrieved sequences (a), (b).

Extensions of STL with metric qualifiers to take into account distances, speeds and the like, allow more fitting assertions on visual data and support more precise reasoning on space and time. An extended version of STL referred to as XSTL (*eXtended Spatio-Temporal Logic*) has been developed and expounded by Del Bimbo and Vicario (1993).

REFERENCES

Allen, J.F. (1983) Maintaining knowledge about temporal intervals. *Communications of the ACM* **26**, 11.

Arndt, T. and Chang, S.-K. (1989) Image sequence compression by iconic indexing. *IEEE VL'89 Workshop on Visual Languages*, Rome, Italy.

Binaghi, E., Gagliardi, I. and Schettini, R. (1992) Indexing and fuzzy logic-based retrieval of color images. In: *IFIP Transactions A-7, Visual Database Systems II*, (Eds Knuth and Wegner), Elsevier, Amsterdam, The Netherlands.

Chang, S.-K. and Jungert, E. (1991) Pictorial data management based upon the theory of symbolic projections. *Journal of Visual Languages and Computing* **2**, 2.

Chang, S.-K. and Liu, S.L. (1984) Picture indexing and abstraction techniques for pictorial databases. *IEEE Trans. on Pattern Analysis and Machine Intelligence* **PAMI-6**, 4.

Chang, S.-K., Shi, Q.Y. and Yan, C.W. (1987) Iconic indexing by 2-D strings. *IEEE Trans. on Pattern Analysis and Machine Intelligence* **PAMI-9**, 3.

Chang, S.-K., Yan, C.W., Dimitroff, D.C. and Arndt, T. (1988) An intelligent image database system. *IEEE Trans. Software Engineering* **14**, 5.

Costagliola, G., Tortora, G. and Arndt, T. (1992) A unifying approach to iconic indexing for 2D and 3D scenes. *IEEE Trans. on Knowledge and Data Engineering* **4**, 3.

Davis, M. (1993) Media streams, an iconic visual language for video annotation. *Telektronik* **4**. Also appeared in reduced version in: *Proc. IEEE VL'93 Workshop on Visual Languages*, Bergen, Norway.

Del Bimbo, A. and Vicario, E. (1993) A logical framework for spatio-temporal indexing of image sequences. *Proc. Workshop on Spatial Reasoning*, Bergen, Norway. To appear also in: *Spatial reasoning* (Eds S.K. Chang and E. Jungert). Plenum Press, New York.

Del Bimbo, M., Campanai, M. and Nesi, P. (1993a) A three-dimensional iconic environment for image database querying. *IEEE Trans. Software Engineering* **SE-19**, 10.

Del Bimbo, A., Vicario, E. and Zingoni, D. (1993b) Symbolic description and visual querying of image sequences using spatio-temporal logic. To appear on *IEEE Trans. on Knowledge and Data Engineering*. Also reduced version in: *Proc. IEEE VL'93 Workshop on Visual Languages*, Bergen, Norway.

Del Bimbo, A., Vicario, E. and Zingoni, D. (1994a) A spatial logic for symbolic description of image contents. *Journal of Visual Languages and Computing* **5**, 1994.

Del Bimbo, A., Pala, P. and Santini, S. (1994b) Visual image retrieval by elastic deformation of object shapes. *Proc. IEEE VL'94 Int. Symp. on Visual Languages*, St Louis, Missouri.

Hirata, K. and Kato, T. (1992) Query by visual example: Content-based image retrieval. In: *Advances in database technology* – Proc. EDBT'92, (Eds A. Pirotte, C. Delobel and G. Gottlob), Lecture Notes in Computer Science, 580, Springer Verlag, Berlin, Germany.

Jungert, E. (1992) The observer's point of view, an extension of symbolic projections. *Proc. Int. Conf. on Theories and Methods of Spatio-Temporal Reasoning in Geographic Space*, Pisa, Lecture Notes in Computer Science, Springer Verlag, Berlin, Germany.

Lee, S. and Hsu, F. (1992) Spatial reasoning and similarity retrieval of images using 2D C-string knowledge representation. *Pattern Recognition* **25**, 3.

Lee, S., Shan, M.K. and Yang, W.P. (1989) Similarity retrieval of iconic image database. *Pattern Recognition* **22**, 6.

Niblack, W. *et al.* (1993) The QBIC project: Querying images by content using color, texture and shape. *Research Report 9203*, IBM Res. Div Almaden Res Center.

Tanimoto, S.L. (1976) An iconic/symbolic data structuring scheme. In: *Pattern recognition and artificial intelligence* (Ed. C.H. Chen). Academic Press, New York.

Walter, I.M., Sturm, R. and Lockemann, P.C. (1992) A semantic network based deductive database system for image sequence evaluation. In: *Visual database systems II* (Eds) Knuth and Wegner). Amsterdam, The Netherlands.

13

Visual Languages for
Tele-action Objects

13.1 INTRODUCTION

Visual languages are receiving increased attention as we move into the multi-media age, because the common user needs new ways to deal with multimedia information directly (S.K. Chang *et al.*, 1995). A graphical user interface (GUI) can display only a limited number of icons simultaneously without cluttering the screen. Moreover, the GUI icons are usually predefined. Visual language systems let the user introduce new icons, and compose icons into visual sentences with different meanings, thus overcoming the GUI's limitations.

In this paper, we extend the definition of visual languages, to develop visual languages for tele-action objects. A *Tele-Action Object* or *TAO* is a multimedia object with associated hypergraph structure and knowledge structure (H. Chang *et al.*, 1995a). The user can create and modify the private knowledge of a tele-action object, so that the tele-action object will automatically react to certain events to pre-perform operations for generating timely response, improving operational efficiency and maintaining consistency. Moreover, a tele-action object also possesses a hypergraph structure, supporting the effective presentation and efficient communication of multimedia information.

For example, a multimedia message is a complex object composed of

ARTIFICIAL VISION
ISBN 0-12-444816-X
Copyright © 1997 Academic Press Ltd
All rights of reproduction in any form reserved

primitive multimedia objects such as text, image, graphics, video and audio objects. Thus the multimedia message possesses a hypergraph structure. If we want the multimedia message to be able to react to certain events, such as alerting the sender when the message has not been viewed by the receiver after three days, then we need to attach private knowledge to the multimedia message (H. Chang *et al.*, 1995b). Therefore, an intelligent multimedia message is a TAO. By incorporating hypergraph and knowledge structures, the tele-action objects can facilitate the development of multimedia applications.

To manipulate TAOs, we can adopt the traditional approach to provide commands to define, insert, delete, modify, and update TAOs. However, since TAOs are multimedia objects, it is more natural to provide a direct means of manipulating them. This poses a challenge to generalize the conceptual framework of visual languages to multidimensional languages. Since multimedia objects are inherently dynamic, such multidimensional languages must be able to deal with dynamic multimedia information.

This paper is organized as follows. In Section 13.2, we describe the generalized icons for tele-action objects. The generalized icons are combined using spatial and temporal operators, which are described in Section 13.3. Section 13.4 deals with the representation of meaning for generalized icons. In Section 13.5, the definition of visual languages is extended to multidimensional languages. We place special emphasis on dynamic multidimensional languages because of their characteristics. Examples are presented in Section 13.6 to illustrate dynamic visual sentences. Section 13.7 presents active index as the knowledge structure of TAO and give a precise definition of TAO. In Section 13.8, we discuss the design of dynamic visual language for the BookMan of a virtual library. Section 13.9 gives some concluding remarks.

13.2 GENERALIZED ICONS

A visual language is a pictorial representation of conceptual entities and operations and is essentially a tool through which users compose iconic, or visual, sentences. Compilers for visual languages must interpret visual sentences and translate them into a form that leads to the execution of the intended task (Chang, 1990). This process is not straightforward. The compiler cannot determine the meaning of the visual sentence simply by looking at the icons. It must also consider the context of the sentence, how the objects relate to one another. Keeping the user's intent and the machine's interpretation the same is one of the most important tasks of a visual language (Crimi *et al.*, 1990).

A visual sentence is a spatial arrangement of object icons and/or process icons that usually describes a complex conceptual entity or a sequence of operations. Object icons represent conceptual entities or groups of object icons that are arranged in a particular way. Process icons denote operations and are usually context dependent. Figure 13.1 shows two visual sentences consisting of horizontally arranged icons, with a dialog_box overlaid on them. The fish tank in Figure 13.2 is a different kind of visual sentence where the fish are object icons and the cat is a process icon (see Section 13.6). The diagram in Figure 13.3 is yet another kind of visual sentence whose icons are connected by directed lines.

From the above introduction, we can see that visual languages are built upon primitives called icons. Icons usually refer to the physical image of an object. However, we also want to deal with different media types, therefore we prefer to

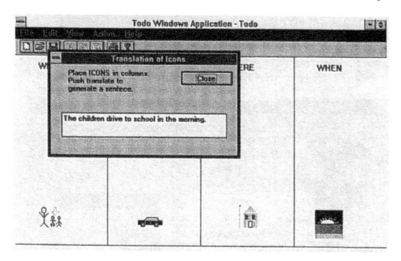

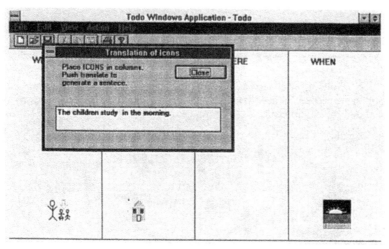

Figure 13.1 Location-sensitive visual sentences (a) and (b) illustrate the change of meaning of the 'school' icon. Such visual sentences can be used to specify to-do items in TimeMan.

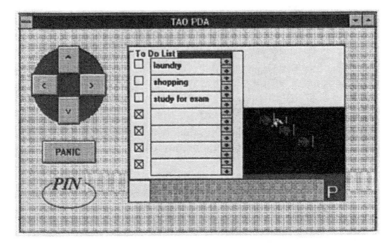

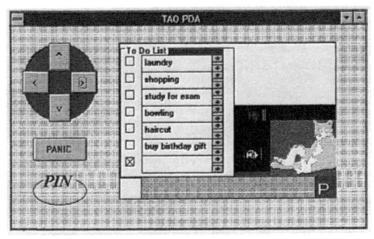

Figure 13.2 Content-sensitive visual sentences (a) and (b) show the fish tank and cat metaphor in TimeMan. When the to-do list grows too long, the fish tank is over-populated and the cat appears.

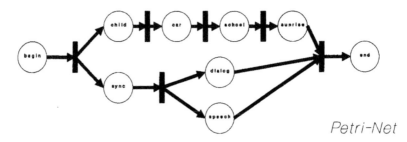

Figure 13.3 A time-sensitive visual sentence is the Petri net controlling the multimedia presentation for the visual sentence of Figure 13.1(a).

use the term *generalized icons*, or *gicons*, for such primitives. Formally, a generalized icon is defined to be $x = (x_m, x_p)$ where x_m is the logical part (the meaning) and x_p is the physical part (Chang, 1987). It is natural to use generalized icons for multimedia TAOs, where we replace the physical part x_p by other media types:

Icon: (x_m, x_i) where x_i is an image
Ticon: (x_m, x_t) where x_t is text (ticon can be regarded as a subtype of icon)
Earcon: (x_m, x_e) where x_e is sound
Micon: (x_m, x_s) where x_s is a sequence of icon images (motion icon)
Vicon: (x_m, x_v) where x_v is a video clip (video icon).

For a generalized icon, the physical part x_p includes an attribute *gicon type* which specifies the media type such as image, text, sound, motion graphics or video, and x_m is the detailed specification of meaning (see Section 13.4).

The *materialization* and *dematerialization* of generalized icons are defined below. In materialization, we obtain the image (or sound, image sequence, video, etc.) from the meaning: $\text{MAT}(x_m) = x_i$ or x_t or x_e or x_s or x_v. Conversely, in dematerialization or interpretation, from the image (or sound, image sequence, video, etc.) we derive the meaning: $\text{DMA}(x_i)$ or $\text{DMA}(x_t)$ or $\text{DMA}(x_e)$ or $\text{DMA}(x_s)$ or $\text{DMA}(x_v) = x_m$.

13.3 SPATIAL TEMPORAL OPERATORS

Icons can be combined using icon *operators*, and the general form of binary operations can be expressed as x_1 *op* $x_2 = x_3$, where the two icons x_1 and x_2 are combined into x_3 using operator *op*. The operator *op* $= (op_m, op_p)$, where op_m is the logical operator, and op_p is the physical operator. Using this expanded notation, we can write (x_{m1}, x_{p1}) *op* $(x_{m2}, x_{p2}) = ((x_{m1}\ op_m\ x_{m2}), (x_{p1}\ op_i\ x_{p2}))$. In other words, the meaning part x_{m1} and x_{m2} are combined using the logical operator op_m, and the physical part x_{p1} and x_{p2} are combined using the physical operator op_p. The representation of meaning will be discussed in Section 13.4.

For multimedia TAOs, we can define earcon operators, micron operators, vicon operators, etc. as follows:

The icon operator $op = (op_m, op_i)$, such as *ver, hor, ovl, con, surround, edge_to_edge*, etc.
The ticon operator $op = (op_m, op_t)$, such as *text_merge*, etc.
The earcon operator $op = (op_m, op_e)$, such as *fade_in, fade_out*, etc.
The micon operator $op = (op_m, op_s)$, such as *zoom_in, zoom_out*, etc.
The vicon operator $op = (op_m, op_v)$, such as *montage*, etc.

One important point to be emphasized is that the icon operators can be *visible* or *invisible*. For example, excluding the dialog_box, the visual sentence shown in

Figure 13.1(a) is the horizontal combination of four icons. Therefore, it can be expressed as:

   ((CHILDREN *hor* CAR) *hor* SCHOOL_HOUSE) *hor* SUNRISE

where *hor* is an invisible operator. But if we look at Figure 13.2, the CAT icon is a visible operator denoting a process to be applied to the fish in the fish tank.

   Visible and invisible operators can be co-present in one visual sentence. For example, two-dimensional mathematical expression can be described as an icon system with visible icon operators, such as +, −, etc. But there are also invisible icon operators such as horizontal combination (*hor*). Therefore, when we write a mathematical expression 1 + 2, we are actually writing an expression,

$$1 \ hor + hor \ 2$$

where + is a visible operator and *hor* is an invisible operator. This is an important characteristic of icon operators which may cause problems in interpreting visual sentences.

   The four most useful domain-independent icon operators are vertical combination (*ver*), horizontal combination (*hor*), overlay (*ovl*), and connect (*con*). In the operator set $OP = \{ver, hor, ovl, con\}$, *ver*, *hor* and *ovl* are usually invisible, and *con* is usually visible as a connecting line.

   In general, we can have both visible icon operators and invisible icon operators. Process icons are usually visible icon operators. An icon operator, whether visible or invisible, have two characteristics:

a)   the operator divides the space into regions;
b)   the operator takes object icons as operands.

   In the materialization of icon operators, a visible icon operator materializes as a visible image, but an invisible icon operator materializes as an empty image (sound, image sequence, video clip). Conversely, in the dematerialization of icon operators, a visible image can be recognized as a visible icon operator, but the empty image may be recognized as an invisible icon operator. It is the second case that can cause problem to the visual language parser, because we may have to postulate the existence of invisible icon operators for the parser to succeed in parsing a visual sentence.

   A visible icon operator can be regarded as a process icon which plays the dual role of an icon operator. But an invisible icon operator, when dematerialized, can become (a) an operator, (b) an object icon, or (c) a process icon.

   We can regard an image as a materialized visual sentence. Therefore, when we dematerialize the image to uncover the icon operators, these icon operators may trigger actions, by causing a division of the space into regions, so that each sub-image in a region can be dematerialized and processed separately.

   The invisible icon operators such as *ver*, *hor* and *ovl* are *spatial operators*. As shown in Table 13.1 such spatial operators can apply only to icons. The spatial composition of two icons is a complex icon. The spatial composition of icon

Table 13.1   Spatial operator.

Spatial operator	Icon	Micon	Earcon	Vicon
Icon	Icon	–	–	–
Micon	–	–	–	–
Earcon	–	–	–	–
Vicon	–	–	–	–

with earcon, micon or vicon does not yield generalized icons, but can be considered as visual sentences or multidimensional sentences (see Section 13.5).

The *temporal operators* are also usually treated as invisible operators. The common temporal operators for earcon, micon and vicon are (Allen, 1983)

co_start, co_end, overlap, equal, before, meet, during, etc.

For the temporal operators, we can construct Table 13.2.

As shown in Table 13.2, a micon can be constructed by the temporal composition of two icons. A micon followed in time by another icon is still a micon. It is interesting to note that the temporal composition of micon and earcon yields vicon.

A spatial/temporal operator *op* has the following attributes: *op_name, op_type, visibility, audibility, narg*, where *op_name* is the name of the operator, *op_type* is either spatial or temporal, *visibility* is either visible or invisible, *audibility* is either audible or inaudible, and *narg* is the number of operands (arguments) of the operator.

13.4 REPRESENTATION OF MEANING FOR GENERALIZED ICONS

To represent the meaning part x_m of the generalized icon (x_m, x_p), a frame (Chang *et al.*, 1994a) or a conceptual graph (Chang *et al.*, 1989) can be employed. Both are appropriate representations of meaning, and can be transformed into one another. For example, the SCHOOL_HOUSE icon in Figure 13.1(a) can be represented by the following frame:

Table 13.2   Temporal operator.

Temporal operator	Icon	Micon	Earcon	Vicon
Icon	Micon	Micon	Vicon	–
Micon	Micon	Micon	Vicon	–
Earcon	Vicon	Vicon	Earcon	–
Vicon	–	–	–	Vicon

Icon SCHOOL_HOUSE
WHO: nil
DO: study
WHERE: school
WHEN: nil

In other words, the SCHOOL_HOUSE icon has the meaning 'study' if it is in the DO location, or the meaning 'school' in the WHERE location. Its meaning is 'nil' if it is in the WHO or WHEN location. An equivalent linearized conceptual graph is as follows:

[Icon = SCHOOL_HOUSE]
−(sub)→   [WHO = nil]
−(verb)→   [DO = study]
−(loc)→   [WHERE = school]
−(time)→   [WHEN = nil]

The meaning of a composite icon can be derived from the constituent icons, if we have the appropriate inference rules to combine the meanings of the constituent icons. Chang *et al.* (1994a) applied the Conceptual Dependency (CD) theory to develop inference rules to combine frames. Chang *et al.* (1989) adopted conceptual operators to combine conceptual graphs. As a simple example, the merging of the frames for the four icons in the visual sentence shown in Figure 13.1(a) will yield the frame:

Visual_Sentence vs_1
WHO: children
DO: drive
WHERE: school
WHEN: morning

The above frame can be derived by merging the frames of the four icons using the following rule: *the ith slot gets the value of the corresponding slot of the ith icon.* Thus the first slot with slot_name WHO gets the value 'children' from the corresponding slot of the first icon CHILDREN, the second slot with slot_name DO gets the value 'drive' from the corresponding slot of the second icon CAR, etc.

13.5 THE SYNTACTIC ASPECT OF MULTIDIMENSIONAL LANGUAGE FOR TELE-ACTION OBJECTS

As mentioned in Section 13.1, with visual languages as the basis, we can define multidimensional languages that can deal with multimedia information more effectively and are capable of handling temporal processes. In this section, we will start with the formal definition of visual languages, and then extend to multidimensional languages whose relational operators can include both spatial and temporal relational operators.

A visual grammar G describes how to generate visual sentences:

$G = (N, X, s, R)$ where
N is the set of nonterminals
X is the set of terminals (icons)
s is the start symbol
R is the set of production rules.

A visual sentence u is generated by applying production rules from G,

$$s \overset{*}{\underset{G}{\Rightarrow}} u.$$

A visual language $L(G)$ is the set of visual sentences generated by a visual grammar G. A visual sentence u is transformed by applying a materialization operator MAT, into an image MAT(u) in the physical space. The MAT is a mapping from X^* into V_P, the physical space of images.

Within the family of visual grammars, the relational visual grammars are of special interest. A relational visual grammar G describes how to generate visual sentences:

$G = (N, X, OP, s, R)$ where
N is the set of nonterminals
X is the set of terminals (icons)
OP is the set of relational operators
s is the start symbol
R is the set of production rules whose right-hand sides are expressions involving relational operators.

We can define the relational visual language hierarchy, where **C1**, **C2** and **C3** are the constraints:

(1) *Positional grammar* with rules of the form $x := a \; op \; b$
C1: Relational operator can be defined between any n symbols.
C2: The number n must be 1 or 2, and the symbols must be adjacent in the production rule.
C3: The evaluation rule for this operator has the characteristics that if we know the left-string, we can predict where to find the next symbol (Costagliola and Chang, 1994).
Parser: parser is positional parser.

(2) *Relational grammar* with rules of the form $x := a \; op \; b$
C1: Relational operator can be defined between any n symbols.
C2: The number n must be 1 or 2, and the symbols must be adjacent in the production rule.
Parser: parser is relational parser.

Comment: the difference between positional grammar and relational grammar is how the *op* is evaluated. If it is by a relational predicate, it is relational grammar. If it is by a positional evaluation rule, it is positional grammar. The fuzzy parser described by Chang (1990) is a relational parser.

(3)   *General relational grammar* with rules of the form $x := (op\ a_1 a_2 \ldots a_n)$

C1:   Relational operator can be defined between any *n* symbols.

Parser: parser is general relational parser.

Comment: the general parser described by Chang (1990) is a general relational parser. Picture Layout Grammar (Golin, 1991) and the Web Grammar (Rosenfeld and Migram, 1972) are variations of general relational grammars.

To extend the above definitions so that we can describe multidimensional languages, we define a relational multidimensional grammar *G* which describes how to generate multidimensional sentences:

$G = (N, X, OP, s, R)$ where
N is the set of nonterminals
X is the set of terminals (icons, earcons, micons, ticons and vicons)
OP is the set of spatial/temporal relational operators
s is the start symbol
R is the set of production rules where the right-hand side must be an expression involving relational operators.

More restrictions can be added to create subsets of rules for icons, earcons, micons and vicons. For each subset, we can define special gicon operators such as: (earcon operators) *fade_in, fade_out*, (micon operators) *zoom_in, zoom_out*, (vicon operators) *montage*, etc. These special gicon operators support the combination of various types of gicons, so that the multidimensional language can fully utilize all the multimedia types.

Informally, a multidimensional sentence is the spatial/temporal composition of generalized icons, and a multidimensional language is a set of multidimensional sentences.

As an example, the multidimensional sentence to generate the visual sentence of Figure 13.1(a) and simultaneously produce a synthesized speech 'The children drive to school in the morning' when the dialog_box is created, has the following syntactic structure:

(DIALOG_BOX *co_start* SPEECH) *ver* (((CHILDREN *hor* CAR) *hor* SCHOOL_HOUSE) *hor* SUNRISE)

As shown in Figure 13.4, the syntactic structure for this multidimensional sentence is essentially a tree with additional temporal operators (such as *co_start*) and spatial operators (such as *hor* and *ver*) indicated by dotted lines. These operators specify the spatial/temporal constraints. Some operators may have more than two operands (for example, the *co_start* of audio, image and text), therefore in general the syntactic structure is a hypergraph.

Figure 13.3 illustrates the Petri-net representation of the dynamics of the multidimensional sentence. We can define transformation rules for the spatial/temporal operators, so that the hypergraph shown in Figure 13.4 can be transformed into a generalized Petri net shown in Figure 13.3 which can be executed dynamically by the multimedia presentation manager to create a multimedia presentation (Lin *et al.*, 1996).

As illustrated by this example, multidimensional languages are inherently dynamic, because many media types (audio, image sequence, video, etc.) are necessarily time-varying. Therefore, the dynamic nature of multidimensional languages should be fully exploited. This leads to the notion of dynamic multidimensional languages.

13.6 THE DYNAMIC ASPECT OF MULTIDIMENSIONAL LANGUAGE FOR TELE-ACTION OBJECTS

For dynamic multidimensional languages, the user is not merely shown a static sentence, but a dynamic one. Since the multidimensional sentence is changing dynamically from one sentence to another sentence, it is called a dynamic multidimensional sentence.

A dynamic multidimensional sentence is defined as $vs = vs_1 vs_2 \ldots vs_n$, where each vs_i is a sentence (the simplest being an icon), and the meaning d_n of the dynamic multidimensional sentence $vs_1 \ldots vs_n$ is computed recursively as follows:

$$d_n = f(d_{n-1}, r(vs_{n-1}, vs_n)), \quad n > 1.$$

In other words, given the meaning d_{n-1} of the dynamic multidimensional sentence $vs_1 \ldots vs_{n-1}$, and the relationship r between vs_{n-1} and vs_n, we can compute the new meaning d_n of the dynamic multidimensional sentence $vs_1 \ldots vs_n$ using the function f.

The dynamic structure of a multidimensional sentence determines how the sentence acquires meaning, i.e., it is also the knowledge structure of the sentence. This knowledge structure can be specified by the active index (see Section 13.7).

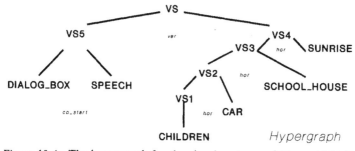

Figure 13.4 The hypergraph for the visual sentence of Figure 13.1(a).

We can describe location-sensitive icons as dynamic multidimensional sentences. In other words, we can first present an example of $vs_1 \ldots vs_n$ where each vs_i is an icon. The background window can also be regarded as an icon. Then, we can specify d_1 (the meaning or description of the first visual icon vs_1), r (the relationship function), and f (the recursive function to compute new meaning), to derive the meaning of the full sentence.

Figure 13.1 shows the *location-sensitive visual sentences*. As illustrated in Figure 13.1, the SCHOOL_HOUSE icon can mean 'school' or 'study' depending upon its location on the screen. The visual sentence in Figure 13.1(a) consisting of the four icons is translated into 'The children drive to school in the morning', and the visual sentence in Figure 13.1(b) consisting of the rearranged four icons correspond to 'The children study in the morning'.

Since there are four regions of the screen, the SCHOOL_HOUSE icon can take on four different meanings. We can define a location-sensitive DMA function as follows: DMA(SCHOOL_HOUSE, 1) = 'nil', DMA(SCHOOL_HOUSE, 2) = 'study', DMA(SCHOOL_HOUSE, 3) = 'school' and DMA(SCHOOL_HOUSE, 4) = 'nil'. In general, the meaning of an icon x in the k region is determined by DMA(x, k).

Suppose the last icon in the visual sentence vs_n is x_n. If we write the visual sentence from left to right, the location of x_n will be in the nth region. Therefore, $r(vs_{n-1}, vs_n) = \text{DMA}(x_n, n)$. Finally, to find the meaning of the dynamic visual sentence, we define $f(d_{n-1}, \text{DMA}(x_n, n))$ to be the concatenation of d_{n-1} and DMA(x_n, n). Here we regard the meaning of the dynamic visual sentence as the concatenated linear sentence produced by the parser, as illustrated in Figure 13.1.

Similarly, context-sensitive icons are also dynamic multidimensional sentences. We can also use Figure 13.1 to illustrate *context-sensitive visual sentences* where the meaning of visual sentences depends upon its context. The example shows a fixed syntax to compose a visual sentence. The window can be regarded as the context of the visual sentence. Depending upon the context, the same visual sentence shown in Figure 13.1(b) can either be a fact 'The children study in the morning', or a query 'What does the children study in the morning?'.

Figure 13.2 illustrates *context-sensitive visual sentences* where the behavior of *TimeMan*, the time management personal digital assistant, depends upon the contents of the fish tank. Figure 13.2 shows the screen design of the *TimeMan*, where the to-do list is visualized as a fish tank and the fish represent items to do. Each time the user adds a new to-do item, a message is sent to the TAO$_{\text{fish}}$. The TAO$_{\text{fish}}$ creates a fish (Figure 13.2(a)) and posts message to the TAO$_{\text{cat}}$. When there are too many to-do items (too many fish in the fish tank), or a to-do item is due, the cat materializes and warns the user (Figure 13.2(b)). The TAO$_{\text{cat}}$ can post messages to other TAO$_{\text{cat}}$s, to prefetch relevant documents such as books in a virtual library, retrieve related medical images for a diagnosis, etc. By incorporating other TAOs, we can add private knowledge to the personal digital assistant.

Time-varying icons are dynamic multidimensional sentences. A simple example is the blinking traffic light. If the time-sensitive visual sentences are generalized icons, they are sometimes called *dynamic icons* (Hsia and Ambler, 1988). Figure 13.3 shows the *time-sensitive visual sentences* for the Petri net model. The visual sentence is the Petri net. Whenever the token distribution changes, the meaning changes as well. Therefore, $r(vs_{n-1}, vs_n) = d_{change}$ is a measure of the changes of the Petri net, and $f(d_{n-1}, d_{change})$ gives the new meaning.

The relationship function r and the meaning function f are realized by the active index, where r is the incremental change between the previous visual sentence and the current one, and f is the consequence caused by the incremental change. The active index is the knowledge structure for the visual language.

13.7 THE ACTIVE INDEX AND TELE-ACTION OBJECTS

We will now describe the index cell, which is the fundamental building block of an active index (Chang, 1995). An *index cell* (ic) accepts input messages and performs some actions. It then posts an output message to a group of output index cells. Depending upon the internal state of the index cell and the input messages, the index cell can post different messages to different groups of output index cells. Therefore the connection between an index cell and its output cells is not static, but dynamic.

An index cell can be either *live* or *dead*. If the cell is in a special internal state called the *dead state*, it is considered dead. If the cell is in any other state, it is considered live. The entire collection of index cells, either live or dead, forms the *index cell base* (ICB). This index cell base ICB may consist of infinitely many cells, but the set of live cells is finite and forms the *active index* (IX).

When an index cell posts an output message to a group of output index cells, these output index cells are activated. If an output index cell is in a dead state, it will transit to the initial state and become a live cell, and its timer will be initialized. On the other hand, if the output index cell is already a live cell, its current state will not be affected, but its timer will be re-initialized.

The output index cells, once activated, may or may not accept the posted output message. The first output index cell that accepts the output message will remove this message from the output list of the current cell. (In case of a race, the outcome is nondeterministic.) If no output index cell accepts the posted output message, this message will stay indefinitely in the output list of the current cell.

After its computation, the index cell may remain active (live) or de-activate itself (dead). An index cell may also become dead, if no other index cells (including itself) post messages to it. There is a built-in timer, and the cell will

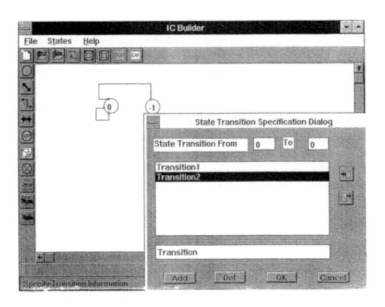

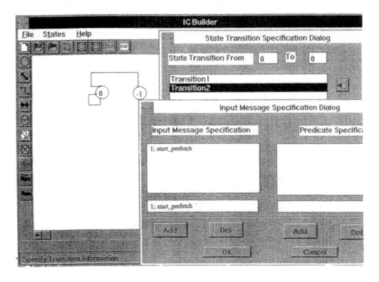

Figure 13.5 The visual specification of the state transitions (a), input message (b), output message and actions (c), for an active index cell of BookMan.

de-activate itself if the remaining time is used up before any message is received. This parameter – the time for the cell to remain live – is re-initialized each time it receives a new message and thus is once more activated.

Although there can be many index cells, these cells may be all similar. For

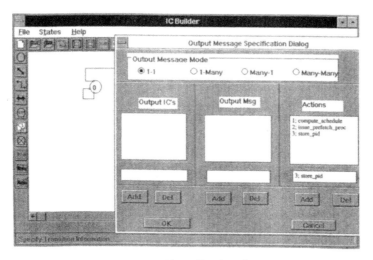

Figure 13.5 Continued.

example, we may want to attach an index cell to a document, so that when a certain feature is detected, a message is sent to the index cell which will perform predetermined actions such as prefetching other documents. If there are ten such documents, then there can be ten such index cells, but they are all similar. These similar index cells can be specified by an *index cell type*, and the individual cells are the instances of the index cell type.

We developed a tool called the *IC_Builder*, which helps the designer construct index cell types using a graphical user interface. The *IC_Builder* can run on any PC with Windows. An example is shown in Figure 13.5.

Figure 13.5(a) illustrates the state transition diagram of an index cell type under construction. This particular index cell type is *prefetch*. This index cell type prefetch is used, together with two other index cell types, in document retrieval (Chang, 1995). Prefetch is responsible for scheduling prefetching, initiating (issuing) prefetching process and killing prefetching process when necessary. The prefetch index cell has two states: state 0 (which is also the initial state) and state −1 (which is the special dead state). The designer uses the graphical tool of the *IC_Builder* to draw this state transition diagram. For example, the designer can click on the fourth icon in the vertical icon menu (the one with a zigzag line) to draw a transition from state 0 to state 0. To avoid too many lines, only one transition line will be drawn, although the designer can specify multiple transitions from state 0 to state 0. In Figure 13.5(a), the designer creates two transitions from state 0 to state 0: transition1 and transition2. Now the details about the highlighted transition2 can be specified.

If the designer clicks on the *input_message* icon located to the right of the state transition specification dialog_box in Figure 13.5(a), *IC_Builder* brings up the input message specification dialog_box shown in Figure 13.5(b), so that the designer can specify the input messages. In this example, the designer specifies message1 (start_prefetch) input message. A predicate can also be specified, and the input message is accepted only if this predicate is evaluated *true*. In this example, there is no predicate, so the input message is always accepted.

Similarly, if the designer clicks on the *output_message* icon located to the right of the state transition specification dialog_box in Figure 13.5(a), *IC_Builder* brings up the output message specification dialog_box shown in Figure 13.5(c), so that the designer can specify the actions, output messages and output ic s. In this example, the designer specifies three actions: action 1 (compute_schedule), action 2 (issue_prefetch_proc) and action 3 (store_pid). (Once a prefetch process is issued, its pid is saved so that the process can be killed later if necessary). There is no output message. Both input and output messages can have parameters, and the output parameters are derived from the input parameters by the index cell.

As described above, a tele-action object has the external appearance of a multidimensional sentence *vs*, with the syntactic structure of a hypergraph derived from *vs* and the dynamic knowledge structure of an active index. The syntactic structure controls its presentation; and the knowledge structure its dynamic behavior.

A tele-action object TAO has the following attributes: *tao_name*, *tao_type*, *p_part*, *link(s)* and *ic(s)*, where *tao_name* is the name of the TAO, *tao_type* is the media type of TAO such as image, text, audio, motion graphics, video or mixed, *p_part* is the physical part of TAO (the actual image, text, audio, motion graphics, video, or a multidimensional sentence for mixed media type), *link* is the link to another TAO and *ic* is the associated index cell.

A TAO can have multiple links and multiple associated index cells. A link has attributes *link_type*, *link_rel* and *link_obj*, where *link_type* is either relational (spatial or temporal) or structural, *link_rel* is either the structural relation *composed_of* or a relational expression involving spatial operators or temporal operators but not both, and *link_obj* is the linked TAO. Whenever the physical part of TAO is changed, such as the modification of the image or the multidimensional sentence, message(s) are sent to the associated ic(s). Sometimes no specific ic is associated with the TAO, i.e., the attribute *ic* is null, in which case message(s) are sent to all ic's in the index cell base ICB, so that those who can respond to the message(s) may be activated.

Formally, when the multidimensional sentence of a TAO is changed from vs_{n-1} to vs_n, a relational function r uses a multidimensional grammar G to analyse the syntactic structures of vs_{n-1} and vs_n, respectively, and maps (vs_{n-1}, vs_n) to a message m in the input message space of the active index. The change from vs_{n-1} to vs_n is due to (a) manual input, (b) external input, or (c) automatic input from the active index. Thus a TAO can react to manual or external inputs, and perform actions and change its own appearance automatically. For example, the manual clicking of the anchor point of the link in the TAO will cause message to be sent to the ic associated with the linked TAO.

13.8 BOOKMAN – AN APPLICATION EXAMPLE

In this section, we describe an application of the tele-action object, which is a VR interface to the virtual library called *BookMan*.

A dynamic visual language for VR such as the virtual library serves as a new query paradigm in a multi-paradigmatic querying system (Chang *et al.*, 1994b). In the *virtual library*, the *physical location* of books can be indicated by marked icons in a graphical presentation of the books stacks of the library. What the user sees on the screen will be the same (with some simplification) as what can be experienced in the real world. The user can select a book by picking it from the shelf, just like in the real world. Therefore, from the viewpoint of dynamic languages, a VR query is a location-sensitive visual sentence.

Figure 13.6(a) illustrates the user can select different query paradigms, such as search by title, author, ISBN and keywords. If the user selects the virtual library search, the user can then navigate in this virtual library as shown in Figure 13.6(b), and the result is rendered as marked objects in a virtual library. If the user switches to a form-based representation by clicking the 'Detailed Record' button in Figure 13.6(b), the result will be rendered as items in a form (Figure 13.6(c)). The user can now use the form to find books of interest, and switch back to the VR query paradigm by clicking the 'VL Location' button in Figure 13.6(c). This example illustrates dynamic querying can be accomplished with greater flexibility by combining the logical paradigms and the VR paradigms.

The admissibility conditions for switching between a *logical paradigm* (such as the form) and a *VR paradigm* (such as the virtual library) is defined as follows. For a logical paradigm, a *VR-admissible query* is an admissible query whose retrieval target object is also an object in VR. For example, the VR for the virtual library contains stacks of books, and a VR-admissible query could be any admissible query about books, because the result of that query can be indicated by marked book icons in the virtual library. Conversely, for a VR paradigm, an *LQ-admissible query* is a VR where there is a single marked VR object that is also a database object, and the marking is achieved by an operation icon such as *similar_to* (find objects similar to this object), *near* (find objects near this object), *above* (find objects above this object), *below* (find objects below this object), and other spatial operators. For example, in the VR for the virtual library, a book marked by the operation icon *similar_to* is LQ-admissible and can be translated into the following query: 'Find all books similar to this book'.

A book, or any document, is also a tele-action object. The user can access the

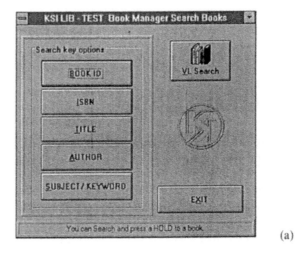

(a)

(b)

Figure 13.6   With the BookMan the user can select search mode (a), browse the virtual library (b) or query by form (c).

book, browse its table of contents, read its abstract, and decide whether to check out this book. Furthermore, if the prefetch mode is selected by the user, the active index will automatically activate the links to access information about related books.

The core of the above described BookMan VR interface for the virtual library has been implemented and is now in daily use by graduate students and faculty members at Knowledge Systems Institute, the graduate school in computers and

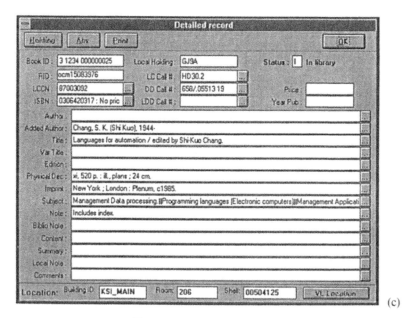

(c)

Figure 13.6 Continued.

management founded by the author. It has become a unique feature of the school. The active index system for the BookMan is currently being implemented, so that optional modes such as prefetching can be supported. In the future we plan to extend BookMan to perform search on the WWW, using an active-index-enhanced Web browser (Chang, 1995).

13.9 CONCLUSION

This paper presented the extension of visual languages into dynamic multi-dimensional languages suitable for the direct manipulation and visualization of tele-action objects. Many applications for multimedia information systems, such as the design of the user interface for PDAs and the virtual library, can be explored.

As demonstrated by BookMan, an important application area for visual languages is visual query systems, which let the common user retrieve information using easy-to-understand visual queries. For example, the user may want to retrieve engineering drawings containing a part that looks like the part depicted in another drawing, and having a signature in the lower-right corner of the document that looks like John Blogg's signature. With visual queries, there is no need to translate the above query into words. The visual sentence allows the direct formulation of such a query with ease. Visual query systems for multimedia databases are currently under active investigation at many universities as well as the industrial laboratories.

Visual languages can also be applied to software specification and documentation. Furthermore, we can program using visual languages. Visual programming languages are now enjoying some initial commercial success. Visual languages are also being successfully applied to augmentative communication systems for the speech impaired (Chang *et al.*, 1994a).

As more work is done in multimedia applications, we expect to see multidimensional language systems, in which visual languages will play an important role, both as a theoretical foundation and as a means to explore new applications.

ACKNOWLEDGEMENTS

This research was supported in part by the National Science Foundation under grant IRI-9224563 and by the Industry Technologies Research Institute of Taiwan. John Howard from my visual languages class contributed Figure 13.1 on location-sensitive visual sentences.

REFERENCES

Allen, J.F. (1983) Maintaining knowledge about temporal intervals. *Communications of the ACM* **26**, 832–843.

Chang, H., Hou, T., Hsu, A. and Chang, S.K. (1995a) Management and applications of tele-action objects. *ACM Multimedia Systems Journal* **3**, 204–216.

Chang, H., Hou, T., Hsu, A. and Chang, S.K. (1995b) Tele-action objects for an active multimedia system. *Proceedings of Second Int'l IEEE Conf. on Multimedia Computing and Systems*, 106–113, Washington, D.C.

Chang, S.K. (1987) Icon semantics – a formal approach to icon system design. *International Journal of Pattern Recognition and Artificial Intelligence* **1**, 103–120.

Chang, S.K. (1990) A visual language compiler for information retrieval by visual reasoning. *IEEE Transactions on Software Engineering*, 1136–1149.

Chang, S.K. (1995) Towards a theory of active index. *Journal of Visual Languages and Computing* **6**, 101–118.

Chang, S.K., Tauber, M.J., Yu, B. and Yu, J.S. (1989) A visual language compiler. *IEEE Transactions on Software Engineering* **15**, 506–525.

Chang, S.K., Orefice, S., Tucci, M. and Polese, G. (1994a) A methodology and interactive environment for iconic language design. *International Journal of Human-Computer Studies* **41**, 683–716.

Chang, S.K., Costabile, M.F. and Levialdi, S. (1994b) Reality bites – progressive querying and result visualization in logical and VR spaces. *Proc. of IEEE Symposium on Visual Languages*, 100–109, St. Louis.

Chang, S.K., Costagliola, G., Pacini, G., Tucci, M., Tortora, G., Yu, B. and Yu, J.S. (1995) Visual language system for user interfaces. *IEEE Software*, 33–44.

Costagliola, G. and Chang, S.K. (1994) Parsing linear pictorial languages by syntax-directed scanning. *Languages of Design* **2**, 223–242.

Crimi, C., Guercio, A., Pacini, G., Tortora, G. and Tucci, M. (1990) Automating visual language generation. *IEEE Transactions on Software Engineering* **16**, 1122–1135.

Golin, E.J. (1991) Parsing visual languages with picture layout grammars. *Journal of Visual Languages and Computing* **2**, 371–393.

Hsia, Y.T. and Ambler, A. (1988) Construction and manipulation of dynamic icons. *1988 Workshop on Visual Languages*, 78–83. Pittsburgh.

Lin, C.C., Xiang, J.X. and Chang, S.K. (1996) Transformation and exchange of multimedia objects in distributed multimedia systems. *ACM Multimedia Systems Journal*, **4**, 12–29.

Rosenfeld, A. and Milgram, D.L. (1972) Web automata and web grammars. *Machine Intelligence* **7**, 307–324.

Index

Printed and bound by CPI Group (UK) Ltd, Croydon, CR0 4YY

03/10/2024

01040418-0018